Lecture Notes in Bioinformatics

T0238181

Subseries of Lecture Notes in Computer Science

Eleazar Eskin   Chris Workman (Eds.)

Regulatory Genomics

RECOMB 2004 International Workshop, RRG 2004
San Diego, CA, USA, March 26-27, 2004
Revised Selected Papers

 Springer

Series Editors

Sorin Istrail, Celera Genomics, Applied Biosystems, Rockville, MD, USA
Pavel Pevzner, University of California, San Diego, CA, USA
Michael Waterman, University of Southern California, Los Angeles, CA, USA

Volume Editors

Eleazar Eskin
University of California, Department of Computer Science and Engineering
San Diego, USA
E-mail: eeskin@cs.ucsd.edu

Chris Workman
University of California, Department of Bioengineering
San Diego, USA
E-mail: chris.workman@gmail.com

Library of Congress Control Number: 2004118137

CR Subject Classification (1998): F.2, G.2, E.1, H.2.8, J.3

ISSN 0302-9743
ISBN 3-540-24456-5 Springer Berlin Heidelberg New York

Springer is a part of Springer Science+Business Media

springeronline.com

© Springer-Verlag Berlin Heidelberg 2005
Printed in Germany

Typesetting: Camera-ready by author, data conversion by Olgun Computergrafik
Printed on acid-free paper   SPIN: 11380795   06/3142   5 4 3 2 1 0

Preface

Research in the field of gene regulation is evolving rapidly in an ever-changing scientific environment. Microarray techniques and comparative genomics have enabled more comprehensive studies of regulatory genomics and are proving to be powerful tools of discovery. The application of chromatin immunoprecipitation and microarrays (chIP-on-chip) to directly study the genomic binding locations of transcription factors has enabled more comprehensive modeling of regulatory networks. In addition, complete genome sequences and the comparison of numerous related species has demonstrated that conservation in non-coding DNA sequences often provides evidence for cis-regulatory binding sites. That said, much is still to be learned about the regulatory networks of these sequenced genomes.

Systematic methods to decipher the regulatory mechanism are also crucial for corroborating these regulatory networks. The core of these methods are the motif discovery algorithms that can help predict cis-regulatory elements. These DNA-motif discovery programs are becoming more sophisticated and are beginning to leverage evidence from comparative genomics (phylogenetic footprinting) and chIP-on-chip studies. How to use these new sources of evidence is an active area of research.

The first RECOMB Regulatory Genomics workshop exceeded the organizers' expectations. More than 130 attendees enjoyed many excellent talks from leading researchers in the field. Ideas were shared during active discussion time between talks and hopefully many collaborations were born. This preceedings contains ten original manuscripts presented by the authors during the workshop. The organizers for the first annual Regulatory Genomics workshop would like to thank all the speakers and participants for their interest and participation in this meeting. The 1st Annual RECOMB Satellite Workshop on Regulatory Genomics would not have been possible without the generous support of UC Discovery and Cal-IT^2.

DATE ♣

Eleazar Eskin
Alkes Price
Ben Raphael
Chris Workman

Organization

Steering Committee

Pierre Baldi	University of California, Irvine
Michael Eisen	Lawrence Berkeley National Lab
Eleazar Eskin (chair)	University of California, San Diego
Pavel Pevzner	University of California, San Diego

Organizing Committee

Eleazar Eskin	University of California, San Diego
Alkes Price	University of California, San Diego
Ben Raphael	University of California, San Diego
Chris Workman	University of California, San Diego

Program Committee

Mathieu Blanchette	McGill University
Julio Collado-Vides	UNAM
Michael Eisen	Lawrence Berkeley National Lab
Mikhail Gelfand	Moscow State University
Sridhar Hannenhalli	University of Pennsylvania
Trey Ideker	University of California, San Diego
Jim Kadonaga	University of California, San Diego
Uri Keich	Cornell University
Manolis Kellis	MIT
Jim Kent	University of California, Santa Cruz
Hao Li	University of California, San Francisco
Dana Pe'er	Harvard University
Yitzhak Pilpel	Weizmann Institute
Mireille Regnier	INRIA
Bing Ren	University of California, San Diego
Marie-France Sagot	INRIA
Eran Segal	Stanford
Ron Shamir	Tel Aviv University
Saurabh Sinha	The Rockefeller University
Rotem Sorek	Compugen
Martin Tompa	University of Washington
Chris Workman	University of California, San Diego
Zohar Yakhini	Agilent
Eric Xing	University of California, Berkeley

Sponsoring Institutions

Industry-University Cooperative Research Program, The UC Discovery Grant
California Institute for Telecommunications and Information Technology, Cal-(IT)^2

Table of Contents

Predicting Genetic Regulatory Response
Using Classification: Yeast Stress Response

Manuel Middendorf, Anshul Kundaje, Chris Wiggins,
Yoav Freund, and Christina Leslie

Columbia University, New York NY 10027, USA
{mjm2007,chris.wiggins}@columbia.edu
{abk2001,freund,cleslie}@cs.columbia.edu

Abstract. We present a novel classification-based algorithm called GeneClass for learning to predict gene regulatory response. Our approach is motivated by the hypothesis that in simple organisms such as *Saccharomyces cerevisiae*, we can learn a decision rule for predicting whether a gene is up- or down-regulated in a particular experiment based on (1) the presence of binding site subsequences ("motifs") in the gene's regulatory region and (2) the expression levels of regulators such as transcription factors in the experiment ("parents"). Thus our learning task integrates two qualitatively different data sources: genome-wide cDNA microarray data across multiple perturbation and mutant experiments along with motif profile data from regulatory sequences. Rather than focusing on the regression task of predicting real-valued gene expression measurements, GeneClass performs the classification task of predicting +1 and -1 labels, corresponding to up- and down-regulation beyond the levels of biological and measurement noise in microarray measurements. GeneClass uses the Adaboost learning algorithm with a margin-based generalization of decision trees called alternating decision trees. In computational experiments based on the Gasch *S. cerevisiae* dataset, we show that the GeneClass method predicts up- and down-regulation on held-out experiments with high accuracy. We explore a range of experimental setups related to environmental stress response, and we retrieve important regulators, binding site motifs, and relationships between regulators and binding sites that are known to be associated to specific stress response pathways. Our method thus provides predictive hypotheses, suggests biological experiments, and provides interpretable insight into the structure of genetic regulatory networks.

Supplementary website: http://www.cs.columbia.edu/compbio/geneclass

1   Introduction

Understanding the underlying mechanisms of gene transcriptional regulation through analysis of high-throughput genomic data has become an important current research area in computational biology. For simpler model organisms such as *S. cerevisiae*, there have been numerous computational approaches that combine gene expression data from microarray experiments and regulatory sequence data to solve different problems in gene regulation: identification of regulatory elements in non-coding DNA [1,2], discovery of co-occurrence of regulatory motifs and combinatorial effects of regulatory

E. Eskin, C. Workman (Eds.): RECOMB 2004 Ws on Regulatory Genomics, LNBI 3318, pp. 1–13, 2005.

molecules [3], and organization of genes that appear to be subject to common regulatory control into "regulatory modules" [4, 5]. Most of the recent studies can be placed broadly in one of three categories: *statistical approaches*, which aim to identify statistically significant regulatory patterns in a dataset [1, 3, 4]; *probabilistic approaches*, which try to discover structure in the dataset as formalized by probabilistic models (often graphical models or Bayesian networks) [5–9]; and *linear network models*, which hope to learn explicit parameterized models for pieces of the regulatory network by fitting to data [10, 11]. While these approaches provide useful exploratory tools that allow the user to generate biological hypotheses about transcriptional regulation, in general, they are not yet adequate for making accurate *predictions* about which genes will be up- or down-regulated in new or held-out experiments. Since these approaches do not emphasize prediction accuracy, it is difficult to directly compare performance of the different algorithms or decide, based on cross-validation experiments, which approach is most likely to generate plausible biological hypotheses for testing in the lab.

In the current work, we present an algorithm called GeneClass that learns a *prediction* function for the regulatory response of genes under different experimental conditions. The inputs to our learning algorithm are the gene-specific regulatory sequences – represented by the set of binding site patterns they contain ("motifs") – and the experiment-specific expression levels of regulators ("parents"). The output is a prediction of the expression state of the regulated gene. Rather than trying to predict a real-valued expression level, we formulate the task as a binary classification problem, that is, we predict only whether the gene is up- or down-regulated. This reduction allows us to exploit modern and effective classification algorithms. GeneClass uses the Adaboost learning algorithm with a margin-based generalization of decision trees called alternating decision trees (ADTs). Boosting, like support vector machines, is a large-margin classification algorithm that performs well for high-dimensional problems. We evaluate the performance of our method by measuring prediction accuracy on held-out microarray experiments, and we achieve very good classification results in this setting. Moreover, we can analyze the learned prediction trees to extract significant features or relationships between features that are associated with accurate generalization rather than just correlations in the training data. In a range of computational experiments for the investigation of environmental stress response in yeast, GeneClass retrieves significant regulators, binding motifs, and motif-regulatory pairs that are known to be associated with specific stress response pathways.

Among recent statistical approaches, the most revelant method related to GeneClass is the REDUCE algorithm of Bussemaker *et al.* [1] for regulatory element discovery. Given gene expression measurements from a single microarray experiment and the regulatory sequence S_g for each gene g represented on the array, REDUCE proposes a linear model for the dependence of log gene expression E_g on presence of regulatory subsequences (or "motifs") $E_g = C + \sum_{\mu \in S_g} F_\mu N_{\mu g}$, where $N_{\mu g}$ is a count of occurrences of regulatory subsequence μ in sequence S_g, and the F_μ are experiment-specific fit parameters. GeneClass generalizes beyond the conditions of a single experiment by using paired (motif$_g$,parent$_e$) features, where the parent variable represents over- or under-expression of a regulator (transcription factor, signaling molecule, or protein kinase) in the experiment e, rather than using motif information alone. Note, however, that GeneClass uses classification rather than regression as in REDUCE.

Restriction to a candidate set of potential parents has also been used in the probabilistic model literature, including in the regression-based work of Segal *et al.* for partitioning target genes into *regulatory modules* for *S. cerevisiae* [5]. Here, each module is a probabilistic regression tree, where internal nodes of the tree correspond to states of regulators and each leaf node prescribes a normal distribution describing the expression of all the module's genes given the regulator conditions. The authors provide some validation on new experiments by establishing that the target gene sets of specific modules do have statistically significant overlap with the set of differentially expressed genes; however, they do not focus on making accurate predictions of differential expression as we do here. Our GeneClass method retains the distinction between regulator ("parent") genes and target ("child") genes, as well as a model that can capture combinatorial relationships among regulators; however, the margin-based GeneClass trees are very different from probabilistic trees. Unlike in [5], we learn from both expression and sequence data, so that the influence of a regulator is mediated through the presence of a regulatory sequence element. We note that in separate work, Segal *et al.* [6] present a probabilistic model for combining promoter sequence data and a large amount of expression data to learn transcriptional modules on a genome-wide level in *S. cerevisiae*, but they do not demonstrate how to use this learned model for predictions of regulatory response.

The current work follows up on our original paper introducing the GeneClass algorithm for prediction of regulatory response [12]. Here, we report additional computational experiments and more detailed biological validation for specific environmental stress responses (Section 4.3). Due to space constraints, we omit some algorithmic details and refer the reader to the earlier presentation and to additional results available at the supplementary website: http://www.cs.columbia.edu/compbio/geneclass.

2   Learning Algorithm

2.1   Adaboost

The underlying classification algorithm that we use is Adaboost, introduced by Freund and Schapire [13], which works by repeatedly applying a simple learning algorithm, called the *weak* or *base* learner, to different weightings of the same training set. For binary prediction problems, Adaboost learns from a training set that consists of pairs $(x_1, y_1), (x_2, y_2), \ldots, (x_m, y_m)$, where x_i corresponds to the features of an example and $y_i \in \{-1, +1\}$ is the binary label to be predicted, and maintains a *weighting* that assigns a non-negative real value w_i to each example (x_i, y_i). On iteration t of the boosting process, the weak learner is applied to the training set with weights $w_1^t, \ldots, w_m^t$ and produces a prediction rule h_t that maps x to $\{0, 1\}$. The rule $h_t(x)$ is required to have a small but significant correlation with the labels y when measured using the current weighting. After the function h_t is generated, the example weights are changed so that the weak predictions $h_t(x)$ and the labels y are decorrelated. The weak learner is then called with the new weights over the training examples and the process repeats. Finally, one takes a linear combination of all the weak prediction rules to obtain a real-valued *strong* prediction function or *prediction score* $F(x)$. The strong prediction rule is given by $\text{sign}(F(x))$:

$$
\begin{aligned}
&F_0(x) \equiv 0 \\
&\text{for } t = 1 \ldots T \\
&\quad w_i^t = \exp(-y_i F_{t-1}(x_i)) \\
&\quad \text{Get } h_t \text{ from } \textit{weak learner} \\
&\quad \alpha_t = \ln\left(\frac{\sum_{i:h_t(x_i)=1, y_i=1} w_i^t}{\sum_{i:h_t(x_i)=1, y_i=-1} w_i^t} \right) \\
&\quad F_{t+1} = F_t + \alpha_t h_t.
\end{aligned}
$$

One can prove that if the weak rules are all slightly correlated with the label, then the strong rule learned by Adaboost will have a very high correlation with the label – in other words, it will predict the label very accurately. Moreover, one often observes that the test error of the strong rule (percentage of mistakes made on new examples) continues to decrease even after the training error (fraction of mistakes made on the training set) reaches zero. This behavior has been related to the concept of a "margin", which is simply the value $yF(x)$ [14]. While $yF(x) > 0$ corresponds to a correct prediction, $yF(x) > a > 0$ corresponds to a *confident* correct prediction, and the confidence increases monotonically with a. Our experiments in this paper demonstrate the correlation between large margins and correct predictions on the test set (see Section 4).

2.2   ADTs for Predicting Regulatory Response

Adaboost is often used with a decision tree learning algorithm as the base learning algorithm. For the problem of predicting regulatory response, we use a form of Adaboost that produces a single tree-based decision rule called an *alternating decision tree* (ADT) [15]. More details on learning ADTs for regulatory response can be found in [12].

Briefly, in our problem setting, we begin with a candidate set of *motifs* μ representing known or putative regulatory element sequence patterns and a candidate set of regulators or *parents* π. For each (gene,experiment) example in our gene expression dataset, we have two sources of feature information relative to the candidate motifs and candidate parent sets: a vector $N_{\mu g}$ of motif counts of occurrences of patterns μ in the regulatory sequence of gene g, and the vector $\pi_e \in \{-1, 0, 1\}$ of expression states for parent genes π in the experiment e. The data representation is depicted in Figure 1 (A).

Figure 1 (B) shows a toy example of an ADT that could be produced by Adaboost in our setting. An ADT consists of alternating levels of *prediction nodes* (ovals) – which contain real-valued contributions to the prediction scores – and *splitter nodes* (rectangles) – which contain true/false conditions. To obtain the prediction score $F(x)$ for a particular example x, we sum the values in all prediction nodes that we can reach along *all* paths down from the root corresponding to yes/no decisions consistent with x.

Splitter nodes in our ADTs depend on decisions based on (motif,parent) pairs. However, instead of splitting on real-valued thresholds of parent expression and integer-valued motif count thresholds, we consider only whether a motif μ is present or not, and only whether a parent π is over-expressed (or under-expressed) in the example. Thus, splitter nodes make boolean decisions based on conditions such as "motif μ is present and regulator π is over-expressed (or under-expressed)". Paths in the learned ADT correspond to conjunctions (AND operations) of these boolean (motif,parent) conditions.

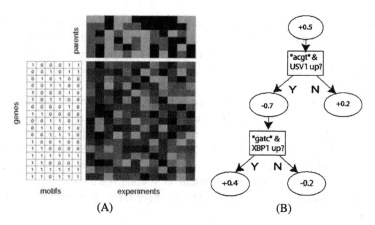

(A)   (B)

Fig. 1. Boosting ADTs for regulatory response prediction. In (A), we show the data presentation for our problem. Every (target gene,experiment) is assigned a label of +1 (up-regulated, in red) or -1 (down-regulated, in green) and represented by the gene's vector of motif counts (only binary values shown here) and the experiment's vector of regulator expression states. A toy example of an ADT is shown in (B)

Full details on selection of the candidate motifs and regulators and discretization into up and down states is given in Section 3.

In terms of Adaboost, each prediction node represents a weak prediction rule, and at every boosting iteration, a new splitter node together with its two prediction nodes is introduced. The splitter node can be attached to any previous prediction node, not only leaf nodes. In general, more important decision rules are added at early iterations. We use this heuristic to analyze the ADTs and identify the most important factors in gene regulatory response.

3   Methods

Dataset: We use the Gasch *et al.* [16] environmental stress response dataset, consisting of cDNA microarray experiments measuring genomic expression in *S. cerevisiae* in response to diverse environmental transitions. There are a total of 6110 genes and 173 experiments in the dataset, with all measurements given as $\log_2$ expression values (fold-change with respect to unstimulated reference expression). We do not perform a zero mean and unit variance normalization over experiments, since we must retain the meaning of the true zero (no fold change).

Motif set: We obtain the 500 bp 5' promoter sequences of all *S. cerevisiae* genes from the Saccharomyces Genome Database (SGD). For each of these sequences, we search for transcription factor (TF) binding sites using the PATCH software licensed by TRANSFAC [17]. The PATCH tool uses a library of known and putative TF binding sites, some of which are represented by position specific scoring matrices and some by consensus patterns, from the TRANSFAC Professional database. A total of 354 binding sites are used after pruning to remove redundant and rare sites.

Parent set: We compile different sets of candidate regulators to study the performance and dependence of our method on the set of regulators. We restrict ourselves to a superset of 475 regulators (consisting of transcription factors, signaling molecules and protein kinases), 466 of which are used in Segal *et al.* [5] and 9 generic (global) regulators obtained from Lee *et al.* [18].

Due to computational limitations on the number of (motif,parent) features we could use in training, we select smaller subsets of regulators based on the following selection criteria. We use 13 high-variance regulators that had a standard deviation (in expression over all experiments) above a cutoff of 1.2. The second subset consists of the 9 global regulators that are present in the Lee *et al.* studies but absent in the candidate list of Segal *et al.* We also include 30 regulators that are found to be top ranking regulators for the 50 clusters identified in Segal *et al.* The union of these three lists gives 53 unique regulators.

Target set and label assignment: We discretize expression values of all genes into three levels representing down-regulation (-1), no change (0) and up-regulation (+1) using cutoffs based on the empirical noise distribution around the baseline (0) calculated from the three replicate unstimulated (time=0) heat-shock experiments [16]. We observe that 95% of the samples in this distribution had expression values between $+1.3$ and -1.3. Thus we use these cutoffs to decide what we define as significantly up-regulated (+1) and down-regulated (-1) beyond the levels of biological and experimental noise in the microarray measurements.

Note that, although we *train* only on those (gene,experiment) pairs which discretize to up- or down-regulated states, we can *test* (make predictions) on every example in a held-out experiment by thresholding on predicted margins. That is, we predict baseline if a prediction has margin below threshold (see Section 4)).

We reduce our target gene list to a set of 1411 genes which include 469 highly variant genes (standard deviation > 1.2 in expression over all experiments) and 1250 genes that are part of the 17 clusters identified by Gasch et al. [16] using hierarchical clustering (eliminating overlaps).

Software: We use the MLJAVA software developed by Freund and Schapire [19] which implements the ADT learning algorithm. We use the text-feature in MLJAVA to take advantage of the sparse motif matrix and use memory more efficiently.

4   Experimental Results

4.1 Cross-Validation Experiments

We first perform cross-validation experiments to evaluate classification performance on held-out experiments. We divide the set of 173 microarray experiments into 10 folds, grouping replicate experiments together to avoid bias, and perform 10-fold cross-validation experiments using boosting with ADTs on all 1411 target genes.

We train the ADTs for 400 boosting iterations, during most of which test-loss decreases continuously. We obtain an accuracy of 88.5% on up- and down-regulated examples averaged over 10 folds (test loss of 11.5%), showing that predicting regulatory response is indeed possible in our framework.

To assess the difficulty of the classification task, we also compare to a baseline method, k-nearest neighbor classification (kNN), where each test example is classified by a vote of its k nearest neighbors in the training set. For a distance function, we optimize the weighted sum of Euclidean distances for motif and parent vectors, trying values of $k < 20$ and both binary or integer representations of the motif data (see [12]). We obtain minimum test-loss of 31.3% at k=19 and with integer motif counts, giving much poorer performance than boosting with ADTs.

Since ADTs output a real-valued prediction score $F(x)$ whose absolute value measures the confidence of the classification, we can predict a baseline label by thresholding on this score, that is, we predict examples to be up- or down-regulated if $F(x) > a$ or $F(x) < -a$ respectively, and to be baseline if $|F(x)| < a$, where $a > 0$. Figure 2 (A) shows expression values versus prediction scores for all examples (up, down, and baseline) from the held-out experiments using 10-fold cross-validation. The plot shows a high correlation between expression and prediction, reminiscent of the actual regression task. Assigning thresholds to expression and prediction values binning the examples into up, down and baseline we obtain the confusion matrix in Figure 2 (B).

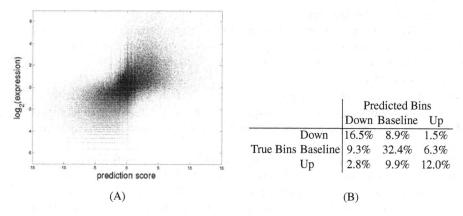

	Predicted Bins		
	Down	Baseline	Up
Down	16.5%	8.9%	1.5%
True Bins Baseline	9.3%	32.4%	6.3%
Up	2.8%	9.9%	12.0%

(A)　　　　　　　　　　　　　　　　　(B)

Fig. 2. True expression values versus prediction scores $F(x)$. The scatter plot (A) shows a high correlation between prediction scores (x-axis) and true log expression values (y-axis) for genes on held-out experiments. The confusion matrix (B) gives truth and predictions for all genes in the held-out experiments, including those expressed at baseline levels. Examples are binned by assigning a threshold $a = \pm 0.5$ to expression and prediction scores

4.2　Extracting Features for Biological Interpretation

We describe below several approaches for extracting important features from the learned ADT model, and we suggest ways to relate these features to the biology of gene regulation.

Extracting significant features: We rank motifs, parents and motif-parent pairs by two main methods. The *iteration score (IS)* of a feature is the boosting iteration during which it first appears in the ADT. This ranking scheme appears to be useful in identifying important motifs and motif-parent pairs (restricting to iteration scores < 50), since features selected at early rounds tend to be most significant. The *abundance score (AS)* of a regulator in the number of nodes in the final tree that include the regulator as the parent in a motif-parent parent. A regulator with a large abundance score will affect a large number of paths through the ADT and hence affect a large number of target genes. If the state of a regulator is changed through stress response or knockout, its predicted effect on target genes will depend on its abundance in the ADT. Note that presence of a strong motif-parent feature does not necessarily imply a direct binding relationship between parent and motif. Such a pair could represent an indirect regulatory relationship or some other kind of predictive correlation, for example, co-occurrence of the true binding site with the motif corresponding to the feature.

"In silico" knock-outs: By removing a candidate from the regulator list and retraining the ADT, we can evaluate whether test loss significantly decreases with omission of the parent and identify other weaker regulators that are also correlated with the labels. We investigate in silico knock-outs in the biologically-motivated experiments described in Section 4.3

4.3   Biological Validation Experiments

We designed the following four training and test sets of selected microarray experiments to study the response to specific types of stress. By comparative analysis of the trees learned from these sets, we find and validate regulators that are associated to regulation programs activated by different stresses. More detailed results can be found on the supplementary website.

Heat-shock and osmolarity stress response: In the first study, we train on heat-shock, osmolarity, heat-shock knockouts, over-expression, amino acid starvation experiments, and we test on stationary phase, simultaneous heat-shock and hypo-osmolarity experiments.

We observe a low test loss of 9.3%, supporting the hypothesis that pathways involved in heat-shock and osmolarity stress appear to be independent and the joint response to both stresses can be predicted easily. This result agrees with the observation by Gasch *et al.* [16] that these two environmental stresses have nearly additive effects on gene expression of environmental stress response (ESR) genes. The high test accuracy also supports the observation by Gasch *et al.* [16] that the response as well as parts of the underlying regulatory mechanisms for stationary phase induction (test set) are similar to that of heat-shock (training set).

The top five high scoring parents (based on AS) were USV1, XBP1, PPT1, GIS1 and TPK1. These regulators are known or believed to play specific important roles in each of the training and test set stresses. Segal *et al.* [5] specifically identify USV1 as an important regulator in stationary phase (test set) and PPT1 to be important in the response to osmolarity stress (training set).

The top ranking motif (based on IS) was the STRE element of MSN2/MSN4, a known regulatory element for a significant number of general stress response genes [16]. The connection of the osmolarity response to the HOG and other MAP kinase pathways is well known. Also, the osmolarity response is strongly related to glycerol metabolism and transport and hence closely associated with gluconeogenesis and glucose metabolism pathways. We find the binding sites of CAT8 (gluconeogensis), GAL4 (galactose metabolism), MIG1 (glucose metabolism and regulator of osmosensitive glycerol uptake) [20], GCN4 (regulator of HOG pathway and amino acid metabolism), HSF1 (heat-shock factor), CHA4 (amino acid catabolism), MET31 (methionine biosynthesis) and RAP1 to have high iteration scores; these regulators are all related to the stress conditions in the training set.

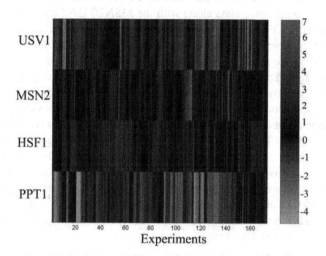

Experiments

Fig. 3. Comparison of expression profiles (173 experiments) of USV1, MSN2, HSF1 and PPT1. The mRNA expression levels of USV1 and PPT1 are informative, with about 50% and 35% of experiments (respectively) showing over 2 fold expression change over wildtype. The espression levels for MSN2 and HSF1 fall mostly in the baseline state, with only about 6% and 5% of experiments (respectively) showing at least 2 fold expression change. While MSN2 and HSF1 are not identified as high scoring parents in the learned trees, their binding sites occur as high scoring motifs

It is interesting to note that while the presence of binding sites of some very important stress factors like MSN2 and HSF1 (heat shock factor) are identified as significant features (high motif iteration score) in the ADT, the mRNA expression levels of these regulators do not seem to be very predictive. HSF1 does not appear as a parent and MSN2 gets low abundance and iteration scores as a parent, despite their importance as heat-shock and general stress response regulators respectively. Similar results are observed in the modules of [5], where HSF1 is not found in any of the regulation programs and MSN2 is found in 3 of the 50 regulation programs but with low significance. If we compare the expression profiles of HSF1, MSN2, USV1 and PPT1, we find that the mRNA levels of MSN2 and HSF1 have small fluctuations (rarely greater than 2 fold change) and fall mostly within the baseline state, while the expression levels of USV1 or PPT1 show much larger variation over many experiments (see Figure 3). It is known that the activity of MSN2 is regulated by TPK1 (a kinase) via cellular localization. TPK1 is identified as an important parent in the ADT (AS) and is found associated with the MSN2 binding site as a motif-parent pair. Thus in this case, where the activity of a regulator is itself regulated post-transcriptionally, we see a clear advantage of using motif data along with mRNA expression data.

USV1 "in silico" knockout for heat-shock and osmolarity stress: Using the same training and test microarrays as in the heat-shock/osmolarity setup, we perform a second study by removing one of the strong regulators, USV1, from the parent set and retraining the ADT. We get a minor but significant increase in test error from 9.3% to 11%.

TPK1 in the upregulated state along with the MSN2/MSN4 binding site is the top scoring feature (IS). TPK1 is also the top scoring regulator based on abundance.

We also study target genes that change label from correct to incorrect due to the removal of USV1. We reason that since these genes require presence of USV1 in the ADT for correct prediction of their regulatory response, they are highly dependent on USV1 activity and provide good candidates for downstream targets of regulatory pathways involving the knocked out parent. We find that 305 target genes change prediction labels. GO annotation enrichment analysis of these target genes reveal the terms cell wall organization and biogenesis, heat-shock protein activity, galactose, acetyl-CoA and chitin metabolism and tRNA processing and cell-growth. These match many of the terms (namely transcription factors, nuclear transport, cell wall and transport sporulation and cAMP pathway, RNA processing, cell cycle, energy, osmotic stress, protein modification and trafficking, cell differentiation) enriched by analyzing GO annotations of genes that changed significantly in a microarray experiment by [5] with stationary phase induced in a USV1 knockout.

Peroxide, superoxide stress, and pleiotropic response to diamide: For the third study, we train on heat-shock, heat-shock knockouts, over-expression, H_2O_2 wild-type and mutant, menadione, DTT experiments, and we test on diamide experiments. Gasch *et al.* [16] consider the diamide response to be a composite of responses to the experiments in the training set. We observe a moderate test loss of 16%, suggesting that this pleiotropic response is more complex than the simpler additive responses to heat-shock and osmolarity.

Although USV1, XBP1 and TPK1 are the top three regulators, we see the emergence of an important parent, YAP1. This factor appears to be dominant in the ADTs of only those setups that include redox related stresses, specifically peroxide and superoxide stresses, in the training sets. It is well documented that YAP1 plays a significant regulatory role in oxidation stresses, and this role correlates well with our findings. We hypothesize that USV1 is not very important for response to diamide based on analysis of the fourth setup below, and we attribute its presence in the ADT to the presence of heat-shock experiments in the training set (based on the first setup). We thus simulate a knockout by removing USV1 from the training set and retraining on the training data. Test loss reduces dramatically from 16% to 9.2%, indicating that USV1's presence in the ADT is detrimental to prediction on diamide response. The ADT for this setup also shows YAP1 associated with its binding motif as an important feature (IS).

Redox and starvation response: In this study, we train on DTT and diamide stresses and response to nitrogen depletion and stationary phase induction. We test on diauxic shift and amino acid starvation experiments. We observe a poor test loss of 27.9%. This poor prediction accuracy could mean that regulatory systems active in experiments in the training set and test set are significantly different. Gasch et al. [16] mention that the starvation responses are quite different from each other and significantly more complex than other stresses (DTT, diamide stress) due to cell-cycle arrest and several secondary effects.

Analysis of the ADT reveals YGL099W (KRE35) as the most abundant regulator. KRE35 also scores among the top 5 candidates in other setups involving redox stresses (such as the third setup above). It could thus be an important regulator for redox related stresses.

We observe that the poor prediction accuracy correlates with the absence of USV1 in the ADT, which is otherwise abundant in the ADTs of all other setups. Since the first three setups show that USV1 is an important regulator for heat-shock response, we add the heat-shock experiments to the training set. As expected, on retraining with this new training set, we get a very significant improvement in prediction accuracy on the same test set (from 27.9% to 16%). This could mean that pathways involved in the heat-shock response are an important component of the more complex response to some starvation responses.

5   Discussion

We have shown that the GeneClass learning algorithm makes accurate predictions of gene regulatory response in yeast over a wide range of experimental conditions. In particular, in experiments related to environmental stress response, examination of the learned GeneClass tree models retrieved important regulators, motifs, and regulator-motif relationships associated with specific stress response pathways. We believe that GeneClass provides a promising new methodology for integrating expression and regulatory sequence data to study transcriptional regulation.

One important next step is to use GeneClass to analyze larger data sets. Since the Gasch dataset that we used here involves only environmental stress response experiments, it is likely that many of the regulatory pathways are not activated and therefore cannot be modeled by analysis of this dataset alone. We hope to extend our studies by incorporating additional and more diverse yeast data sets currently available through resources like NCBI's Gene Expression Omnibus. At the same time, we plan to make improvements in the computational efficiency of the GeneClass software to allow a significant increase the number of parents so that we can identify the possible roles of many additional regulators. In particular, we plan to use using data structures more appropriate for our pairwise interaction features and weighted sampling to reduce the size of the memory required for holding the training data.

A second potential advance would be a more careful treatment of the raw data. While the *ratio* data (perturbation/wild type) gives a natural input variable for our analysis, better signal to noise is likely to be achieved by taking into account the expression levels separately. In further work, we plan to use two-color noise modeling to establish expression-level specific thresholds and thus allow inclusion of more genes whose up- or down-regulated states currently fall within the baseline category. This improvement will allow more training examples and should enable us to accurately predict the response of more subtle target genes.

A third direction for improvement would be to treat parent and child expression levels as continuous (rather than binary) quantities. Similarly, the number of motifs in the regulatory region, rather than merely their presence/absence, could be taken into account. While these refinements could potentially yield more realistic models, it is important that they be represented in a way that is informative for the learning problem and avoids overfitting.

Acknowledgments

AK is supported by NSF EEC-00-88001. CW and MM are partially supported by NSF ECS-0332479 and NIH GM36277. CL and CW are supported by NIH grant LM07276-02, and CL is supported by an Award in Informatics from the PhRMA Foundation.

References

1. Bussemaker, H.J., Li, H., Siggia, E.D.: Regulatory element detection using correlation with expression. Nature Genetics **27** (2001) 167–171
2. Hughes, J.D., Estep, P.W., Tavazoie, S., Church, G.M.: Computational identification of cis-regulatory elements associated with groups of functionally related genes in Saccharomyces cerevisiae. J. Mol. Biol. **296** (2000) 1205–14
3. Pilpel, Y., Sudarsanam, P., Church, G.M.: Identifying regulatory networks by combinatorial analysis of promoter elements. Nature Genetics **2** (2001) 153–159
4. Ihmels, J., Friedlander, G., Bergmann, S., Sarig, O., Ziv, Y., Barkai, N.: Revealing modular organization in the yeast transcriptional network. Nature Genetics **31** (2002) 370–377
5. Segal, E., Shapira, M., Regev, A., Pe'er, D., Botstein, D., Koller, D., Friedman, N.: Module networks: Identifying regulatory modules and their condition specific regulators from gene expression data. Nature Genetics **34** (2003) 166–176
6. Segal, E., Yelensky, R., Koller, D.: Genome-wide discovery of transcriptional modules from DNA sequence and gene expression. Bioinformatics **19** (2003) 273–282
7. Hartemink, A.J., Gifford, D.K., Jaakkola, T.S., Young, R.A.: Using graphical models and genomic expression data to statistically validate models of genetic regulatory networks. Pac. Symp. Biocomp. (2001) 422–33
8. Pe'er, D., Regev, A., Elidan, G., Friedman, N.: Inferring subnetworks from perturbed expression profiles. Proc. of the Ninth International Conf. on Intelligent Systems for Molecular Biology (2001) 215–224
9. Pe'er, D., Regev, V., Tanay, A.: A fast and robust method to infer and characterize and active regulator set for molecular pathways. Proc. of the Tenth International Conf. on Intelligent Systems for Molecular Biology (2002) 258–267
10. Yeung, M., Tegner, J., Collins, J.J.: Reverse engineering gene networks using singular value decomposition and robust regression. Proc. Natl. Acad. Sci. USA **99** (2002) 6163–8
11. D'Haeseleer, P., Wen, X., Fuhrman, S., Somogyi, R.: Linear modeling of mRNA expression levels during CNS development and injury. Pac. Symp. Biocomp. (1999) 41–52
12. Middendorf, M., Kundaje, A., Wiggins, C., Freund, Y., Leslie, C.: Predicting genetic regulatory response using classification. http://www.cs.columbia.edu/compbio/geneclass (2004)
13. Schapire, R.E.: The boosting approach to machine learning: An overview. In: MSRI Workshop on Nonlinear Estimation and Classification. (2002)
14. Schapire, R.E., Freund, Y., Bartlett, P., Lee, W.S.: Boosting the margin: A new explanation for the effectiveness of voting methods. The Annals of Statistics **26** (1998) 1651–1686
15. Freund, Y., Mason, L.: The alternating decision tree learning algorithm. Proc. of the Sixteenth International Conf. on Machine Learning (1999) 124–133

16. Gasch, A.P., Spellman, P.T., Kao, C.M., Carmel-Harel, O., Eisen, M.B., Storz, G., Botstein, D., Brown, P.O.: Genomic expression programs in the response of yeast cells to environmental changes. Mol. Biol. Cell **11** (2000) 4241–4257

17. Wingender, E., Chen, X., Hehl, R., Karas, H., Liebich, I., Matys, V., Meinhardt, T.., Prüss, M., Reuter, I., Schacherer, F.: TRANSFAC: an integrated system for gene expression regulation. Nucleic Acids Research **28** (2000) 316–319

18. Lee, T.I., Rinaldi, N.J., Robert, F., Odom, D.T., Bar-Joseph, Z., Gerber, G.K., Hannett, N.M., Harbison, C.R., Thompson, C.M., Simon, I., Zeitlinger, J., Jennings, E.G., Murray, H.L., Gordon, D.B., Ren, B., Wyrick, J.J., Tagne, J., Volkert, T.L., Fraenkel, E., Gifford, D.K., Young, R.A.: Transcriptional regulatory networks in Saccharomyces cerevisiae. Science **298** (2002) 799–804

19. Schapire, R.E., Singer, Y.: Boostexter: A boosting-based system for text categorization. Machine Learning **39** (2000) 135–168

20. Klein, C.J., Olsson, L., Nielsen, J.: Glucose control in saccharomyces cerevisiae: the role of MIG1 in metabolic functions. Microbiology **144** (1998) 13–24

Detecting Functional Modules
of Transcription Factor Binding Sites
in the Human Genome

Thomas Manke, Christoph Dieterich, and Martin Vingron

Max-Planck-Institute for Molecular Genetics
Ihnestraße 63–73, D-14195 Berlin, Germany
{manke,dieterich,vingron}@molgen.mpg.de

Abstract. This paper presents a method for predicting biologically meaningful modules of transcription factors. For this purpose, we employ the CORG database of conserved transcription factor binding sites. We aim at enhancing the power of *in-silico* binding site predictions by employing three crucial constraints. First, we rely on conserved promoter regions of orthologous genes in human and mouse, second we look for synergistic transcription factor modules which bind upstream regions preferentially together, and finally we restrict our results to those modules, whose genes have a significant functional overlap. Many of our predicted binding sites coincide with known biological facts as is evidenced by a direct comparison with a single large-scale experiment for E2F binding. We also identified known combinations of transcription factors with a functional enrichment in the set of their shared target genes. Several new modules are suggested for experimental investigation. Finally we study the transcription factor network and suggest a classification of transcription factors according to their regulatory power and control.

1   Introduction

Deciphering mechanisms of gene regulation is a major challenge in functional genomics. Bioinformatics supports this task in various ways. Here we report on our approach to find putative transcription factor binding sites, synergistic relations of binding sites and the resulting network of transcription factor interactions.

The detections of transcription factor binding sites (TFBS) has been very successful in yeast, where the intergenic regions are small enough to inspect them for presence of sequence patterns (Tavazoie *et al.* [1]). In mammalia, in contrast, it is notoriously difficult to pinpoint the transcriptional start sites and the search for motifs cannot be focused on a particular region. However, it has been shown that active TFBS are often conserved across species as the corresponding sequence elements in the genome are usually under selective pressure [2]. We took this as our motivation to compile the CORG database of upstream regions conserved in human and mouse and subsequentially identified evolutionary conserved binding sites in both genomes. The procedure has been previously reported in [3].

Based on a gross abstraction of the complex process of protein-DNA binding, the occurrence of TFBSs in upstream regions of genes can be represented as a bipartite

E. Eskin, C. Workman (Eds.): RECOMB 2004 Ws on Regulatory Genomics, LNBI 3318, pp. 14–21, 2005.

graph or a Boolean binding site matrix of transcription factors versus genes. This matrix contains a 1 whenever a particular binding site is found in an upstream region of a gene and 0 otherwise. We deem this computational approach analogous to the technique of chromatin immuno-precipitation. In [4] this technique was used to determine binding sites for E2F-1/4 in intergenic regions of human cell-cycle related genes. As such detailed experimental binding information is not available for the majority of other transcription factors, we utilize the binding predictions based on known sequences patterns in evolutionary conserved promoter regions.

In the following we demonstrate how this information can be used to derive gene modules which are regulated by the same transcription factor (TF) or the same combination of TFs. We evaluate the functional coherence of those modules using the annotations of the Gene Ontology database. In the subsequent section we compare our results directly with a large-scale experimental data set. In Section 3 we identify preferential co-occurrence of binding sites as this is thought to reflect the modular architecture of transcriptional control [5]. We evaluate the functional coherence of the corresponding gene modules and are able to reason on their likely biological role. Finally, in Section 4, we investigate the properties of the transcription factor network.

2   Validation – Comparison with Experiment

We determined putative binding sites in the conserved regions (as defined in [3]) by screening for known motifs, which were taken from the TRANSFAC database [6]. In the case of string representations we used motifs of length greater than 5 nucleotides and accepted only exact matches. This resulted in a total of 529113 binding sites for 384 TFs and 12719 regulated genes in the human genome.

In order to compare our predictions with biological data we choose as reference the experiment from Ren *et al.* [4], where the authors studied the binding of E2F-1 and E2F-4 to the promoter regions of approximately 1200 cell-cycle regulated genes. In our genome-wide analysis we did not select promoters according to their expression profiles, but rather their evolutionary conservation. Therefore we have a much larger set of ≈ 13000 promoters, of which 886 could be mapped to the data set of Ren *et al.* In Table 1 we present the overlap of our predictions with their findings.

To appreciate the resulting overlap it is important to reiterate that both *in-vivo* and *in-silico* binding data are subject to sizeable false-positives error rates. While an experimental binding site could also be the result of indirect binding to the promoter region,

Table 1. In this table we compare the number of E2F-4 and E2F-1+4 target genes with the biological binding data of [4]. The second column denotes the experimentally observed number of bound promoter regions. The other columns give the number of conserved promoters which contain a known E2F binding motif, the overlap with experiment, and the corresponding p-value calculated from the hypergeometric distribution. The total overlap of our conserved promoter regions with the experimental data set is 886.

TF	Ren	CORG	Overlap	P
E2F-4	79	240	43	6.1×10^{-8}
E2F-1+ 4	38	240	26	6.3×10^{-8}

our prediction is equally error prone as many of the detected pattern may not be functional biologically. Error rates from computational prediction algorithms can to some extent be tuned and optimized to meet one's desires. Usually, though, the false positive rates are overwhelming since nucleotide patterns alone are not powerful enough to encompass the complexity of biological binding, and any such description is necessarily not very specific. On the other hand, we accept that some biological binding sites may not be identified, as our prescription (exact matches) could be too restrictive in some cases. Experimental methods may equally miss active binding events, as they also impose a somewhat arbitrary cut-off to separate noise from signal.

We therefore tested whether the observed intersect could be simply a random artefact. This is not the case; more than half (43) of the E2F-4 binding promoters from Ren *et al* are also found by our approach. This is a highly significant overlap ($p \approx 6 \times 10^{-8}$) and more than what has been reported in [7] (28 common binding sites). In anticipation of the subsequent sections, we also tested the combined occurrence of the pair (E2F-1, E2F-4) against a random choice of gene sets with the same size and find an equally significant overlap of 68%.

To continue the functional analysis in analogy with Ref. [4], we studied the distribution of GO-categories in a putative set of E2F-(1,4) regulated genes and compare it to the background of all 12719 genes in the CORG database. We find a significant enrichment (all $p < 10^{-6}$) of genes which are annotated as transcriptional regulators (GO:0006355), development (GO:0007275), DNA-replication (GO:0006260), DNA binding activity (GO:0003677) and damaged DNA binding activity (GO: 0003684). This agrees with known functions of E2F and we take it as an indication that, with exact matches in conserved regions, we are indeed capable of determining a large fraction of biologically relevant TFBS. In the following sections we extend this study to a genome-wide identification of TF modules and the functional enrichment of their target genes.

3   Identification and Annotation of TF Modules

Here we identify TF modules which are over-represented in our data set and to which we can assign a functional role.

3.1   Single Transcription Factors

As a first step we extend the analysis of the previous section to those 168 transcription factors and their regulated gene sets for which there is an exact motif match in *both* the human and mouse conserved regions, i.e. we discard all exact matches which occur in only one of the two organisms. The gene sets were systematically screened for functional enrichment. First, we determined for each set the GO-category with the best overlap (as defined by the smallest value, h_0, of the hypergeometric distribution). Then we calculated a proper p-value, by choosing 1000 random gene sets of the same size and counting how often one observes a value $h \leq h_0$. In doing so, our background model respects the hierarchical dependencies within the gene ontology. We stress that the random selection process is biased proportionally to the size of the conserved regions for each gene. A similar test can be done to assess whether the overall distribution of GO-categories deviates significantly from the background. The same methods were later applied to gene sets which are regulated by combinations of transcription factors as described in the following sections. In Table 2 we present our list of single TFs which show a significant overlap with a functional category as defined by a false positive rate, $p < 0.001$. At this rate we would expect less than one association by mere chance for the 168 gene sets in question.

Table 2. The above list of TF has conserved motif matches in promoters, whose gene sets show significant association with a functional category. In the last column we give the fraction of genes in the set which share the specified function. Only sets with $p < 0.001$ are shown and a few cases have been removed, where the target gene set only contains one or two genes (very specific motif).

TF	GO-category of target genes	overlap
E2F	regulation of transcription	105/543
CP1	regulation of transcription	47/192
NF-κB	chemokine activity	5/30
HrpF	proton transport	6/34
SRF	muscle development	6/34

3.2   Transcription Factor Modules from Biclusters

Eukaryotic cells often utilize a modular promoter architecture to control gene expression under a wide variety of developmental and environmental conditions. The algorithmic challenge is to identify groups of transcription factors, which preferentially bind to certain groups of promoters in a noisy and heterogenous set of binding data. To this end many different clustering algorithms have been developed [8–11]. Here we interprete the binding data as a directed bipartite graph with two classes of nodes (TFs $\rightarrow$ promoters) and implement an algorithm, which was originally suggested in [12] for biclustering of microarray data (conditions $\rightarrow$ genes). This is a greedy algorithm that finds a number of small complete subgraphs (bicliques) and extends them until their weight cannot be improved. For the latter we use the log-odd score of a cluster model versus a randomized graph model with same connectivity. While the weight of a given subgraph can be used to rank bicliques and biclusters for their topological relevance, we postprocessed the high-ranking clusters further, and screened their gene sets against different GO-categories.

Specifically, we identified 1553 distinct heavy modules sharing between 2 and 4 TFs. In Table 3 we list some of the TF-combinations which show a significant functional coherence of the regulated gene set. As before, the false positive rate was calculated by sampling of random gene sets with the same size.

We observe a number of suggestive modules with clear biological function. Many of the TFs with known motifs have been studied because of their role in development and their capacity to regulate other transcription factors. Therefore it comes as no surprise that many of the significant associations are also coherent in those functional categories, only a few of which are shown in Table 3. What is more interesting to observe is how combinations of transcription factors can convey a higher functional coherence. Consider, for example, the set of 34 promoters bound by both CP1 and NF-YA. Individually,

Table 3. This list shows the functional enrichment of selected TF-modules. We stress that it is not simply based on the (large) fractional overlap of the target genes with a given GO-category, but rather on the significant p-value which is smaller than 0.001 in all cases. We have removed some redundancies in cases where a module has significant overlap with several related functional categories.

TF-module	GO-category of target genes	overlap
ATF c-Ets-1 Elk-1	transcription factor activity	7/10
ATF Elk-1 NP-TCII	transcription factor activity	11/18
E2F Elk-1 TFIID	transcription factor activity	9/10
ATF δCREB NF-YA	organogenesis	7/12
ATF AML1a	development	17/37
c-Ets-1 Elk-1 NP-TCII	development	10/17
E2F FOXL1	development	18/35
ATF UBP-1 NP-TCII	retinoic acid receptor activity	3/6
CP1 NF-YA	MHC class II receptor activity	6/34
GABP HrpF	proton transport	6/18

their binding pattern can be found in 258 and 453 conserved promoters, which are by themselves not very specific. However, in combination we find a significant enrichment in genes with "MHC class II receptor activity" and almost half of all target genes are involved in receptor activity, protein targeting and signal transduction. This enhanced specificity is characteristic for all modules shown in Table 3.

In order to derive the biclusters we discarded a number of most abundant transcription factors with unspecific binding patterns (such as PEA3, GATA-2, Sp1 and others). Clearly, some of them are utilized by the cell in many different processes, but their function and specificity is likely to be determined by other (non-sequence) signals.

4 Transcription Factor Circuitry

The ultimate goal of transcriptional control analysis is the identification of regulatory networks in which environmental stimuli propagate through a signal transduction cascade to their respective promoters. The corresponding genes may themselves be transcriptional regulators and often drive complex processes (such as the cell-cycle). Since there is a shortage of large-scale protein interaction data for humans, we can only take a first glimpse at such a network. In particular we studied direct genetic interactions of transcription factors as they are frequently recorded in the CORG database (a prominent example being the one of E2F-1, which binds to its own promoter). An important first question about such networks concerns the hierarchies of regulation and the search for "master" regulators. To address this issue in a quantitative manner we introduce two measures of *regulatory power* (P_t) and *regulatory control* (C_t) for each transcription factor t:

$$P_t = \frac{\text{number of TFs regulated by } t}{\text{total number of genes regulated by } t} \tag{1}$$

$$C_t = \frac{\text{number of TFs regulating } t}{\text{number of conserved basepairs}} \tag{2}$$

Basically, these are conveniently normalized out- and in-degrees in the bipartite graph and they take into account the heterogeneities in the specificity of the TFs and the conserved promoters (i.e. their lengths). These quantities have been further rescaled by their averages, $\bar{P}$ and $\bar{C}$. In Figure 1 we present a view of the TF-network as defined by CORG. Each factor in this coordinate system is placed according to its values C_t and P_t. Not shown are those factors for which the encoding genes show no conservation in their upstream regions and for which we have no regulatory information.

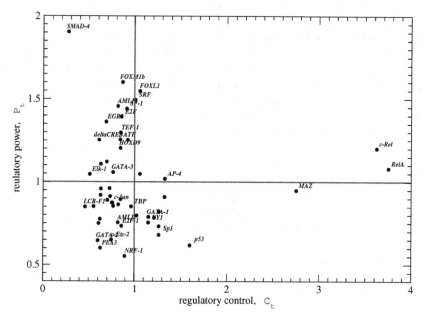

Fig. 1. Network of transcription factors. The nodes are arranged such that transcription factors which regulate many other TFs appear further at the top, and TFs which are regulated by many other TFs appear further to the right. The precise definition of these quantities is given in the main text.

Transcription factors with enhanced binding to other TF-promoters include SMAD-4 and FOXL1. On the other hand, we also observe transcription factors with an apparent depletion of TF-targets, such as p53 and NRF-1. One may speculate that evolution could have discouraged the binding of some transcription factors to the promoter regions of other regulators. Other transcription factors, such as c-Rel and RelA, seem to be tightly regulated as we observe a rather high density of binding sites in their conserved promoter regions.

5   Conclusions

With the CORG database we have created a comprehensive and high-quality database of predicted transcription factor binding sites which opens up the way for further exploratory analysis. In this work we studied the co-occurrences and the modular architecture of regulatory networks based on binding motifs in conserved intergenic regions. In a careful validation study we find 54% of all E2F-4-targets, which were reported experimentally by Ren *et al.* [4]. For the TF-pair E2F-(1,4) the overlap is even higher ($\approx 68\%$). The false positive rate is probably very high, but one cannot properly estimate it as negative experimental evidence is usually not available. Here we assume that biologically meaningful (modular) binding events will stick out of the random background as having a higher degree of co-occurrence than what one would expect for random hits.

In our biclustering approach we focused merely on the presence or absence of a factor in a given promoter region. In several cases one can observe clear hints of functional enrichment and we consider those to be reliable targets for future research as they are endorsed by human-mouse orthology, co-occurring TFBS, and an unusual overlap with GO-categories. In particular the utilization of biclusters helped in establishing functional associations and one would like to reason that this resembles the way in which the cell endows specificity to modules, rather than individual factors.

The validation of modules using functional annotations should be considered a first step only and awaits further analysis. There is much room to supplement this simple measures of coherence with other information, such as co-expression and tissue specificity. Reliability may also be increased by reducing the pattern space to clusters of similar binding site motifs.

Finally, we studied the network of transcription factors and defined two measures to distinguish powerful factors, which regulate many other transcription factors, and those which are themselves heavily regulated. We identified a number of key regulators such as SMAD-4 and FOXL1. Factors under comparatively tight control are c-Rel, RelA and MAZ. A more complete identification of the human regulatory network will depend in large parts on the availability of protein interaction data, which could be naturally included into our framework.

Acknowledgements

We would like to thank Steffen Grossmann for valuable discussions. T.M. acknowledges funding by European Community Contract No. QLRI-CT-2001-00015 for "TEMBLOR" under the specific RTD programme "Quality of Life and Management of Living Resources".

References

1. Tavazoie, S., Hughes, J.D., Campbell, M.J., Cho, R.J., Church, G.M.: Systematic determination of genetic network architecture. Nat Genet **22** (1999) 281–285
2. Duret, L., Bucher, P.: Searching for regulatory elements in human noncoding sequences. Curr Opin Struct Biol **7** (1997) 399–406
3. Dieterich, C., Cusack, B., Wang, H., Rateitschak, K., Krause, A., Vingron, M.: Annotating regulatory dna based on man-mouse genomic comparison. Bioinformatics **18 Suppl 2** (2002) S84–90
4. Ren, B., Cam, H., Takahashi, Y., Volkert, T., Terragni, J., Young, R.A., Dynlacht, B.D.: E2f integrates cell cycle progression with dna repair, replication, and g(2)/m checkpoints. Genes Dev **16** (2002) 245–56

5. Levine, M., Tjian, R.: Transcription regulation and animal diversity. Nature **424** (2003) 147–51
6. Matys, V., Fricke, E., Geffers, R., Gossling, E., Haubrock, M., Hehl, R., Hornischer, K., Karas, D., Kel, A.E., Kel-Margoulis, O.V., Kloos, D.U., Land, S., Lewicki-Potapov, B., Michael, H., Munch, R., Reuter, I., Rotert, S., Saxel, H., Scheer, M., Thiele, S., Wingender, E.: Transfac: transcriptional regulation, from patterns to profiles. Nucleic Acids Res **31** (2003) 374–378
7. Elkon, R., Linhart, C., Sharan, R., Shamir, R., Shiloh, Y.: Genome-wide in silico identification of transcriptional regulators controlling the cell cycle in human cells. Genome Res **13** (2003) 773–80
8. Alon, U., Barkai, N., Notterman, D.A., Gish, K., Ybarra, S., Mack, D., Levine, A.J.: Broad patterns of gene expression revealed by clustering analysis of tumor and normal colon tissues probed by oligonucleotide arrays. Proc Natl Acad Sci U S A **96** (1999) 6745–50
9. Cheng, Y., Church, G.M.: Biclustering of expression data. Proc Int Conf Intell Syst Mol Biol **8** (2000) 93–103
10. Sheng, Q., Moreau, Y., Moor, B.D.: Biclustering microarray data by gibbs sampling. Bioinformatics **19 Suppl 2** (2003) II196–II205
11. Kluger, Y., Basri, R., Chang, J.T., Gerstein, M.: Spectral biclustering of microarray data: coclustering genes and conditions. Genome Res **13** (2003) 703–16
12. Tanay, A., Sharan, R., Shamir, R.: Discovering statistically significant biclusters in gene expression data. Bioinformatics **18 Suppl 1** (2002) S136–44

Fishing for Proteins in the Pacific Northwest

William Krivan

ZymoGenetics, Inc., 1201 Eastlake Avenue East, Seattle, Washington 98102, USA
`krivan@zgi.com`
`http://www.zymogenetics.com`

Abstract. The experimental characterization of novel genes is a tedious and expensive process. While computational gene characterization cannot replace wet lab studies, it has the potential of providing valuable guidance for experimental biologists. We use the novel IL28A,B and IL29 cytokine family to illustrate an approach to the computational identification and characterization of putative transcriptional regulatory regions that utilizes a combination of known and novel techniques. We then apply the approach, which we dubbed Orthocluster, to screen putative regulatory regions of a set of annotated RefSeq mRNA sequences. The encouraging results obtained for known genes motivate the analysis of novel or uncharacterized genes, which is one focus of our current work.

1 IL-28, IL-29 and Their Class II Cytokine Receptor IL-28R

Cytokines play a critical role in modulating the innate and adaptive immune systems. A family of three cytokines, designated interleukin 28A (IL-28A), IL-28B and IL-29, that are distantly related to Type I interferons (IFNs) and the IL-10 family, was identified from human genomic sequence. Like Type I IFNs, IL-28 and IL-29 are induced by viral infection and show antiviral activity [1]. However, unlike all Type I IFNs, IL28 and IL29 signal through a receptor distinct from the Type I interferon receptor.

1.1 Gene Structure

Figure 1 illustrates a comparison of the genomic sequences containing the human (abscissa) and murine (ordinate) IL-28A genes, showing the conservation of the gene structure for both, the coding and non-coding portions of the gene. The conservation of the non-coding regions labeled 1–4 suggests that they might have regulatory function.

1.2 Identification of Putative Regulatory Elements

A search for binding sites of a set of 40 transcription factors (TFs) gave putative sites of TFs playing roles in the transcriptional regulation of interferons and interferon-stimulated genes (AP-1, CREB, GATA, ISRE, NF-AT, and NF-κB).

In Fig. 2, the pairwise sequence alignments of the conserved regions labeled 1 and 2 in Fig. 1 are displayed along with the positions of conserved putative TF binding sites detected at a relative weight matrix threshold of 60%. The set of putative binding sites identified so far is consistent with the human/mouse alignment. This is an example of

E. Eskin, C. Workman (Eds.): RECOMB 2004 Ws on Regulatory Genomics, LNBI 3318, pp. 22–29, 2005.

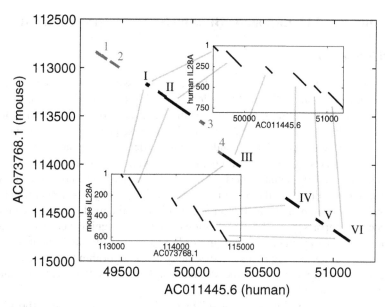

Fig. 1. Gene Structure of Human and Murine IL28A. The six segments shown in black and labeled with roman numerals depict exons. The insets show the alignments of the cDNA sequences with genomic sequence. The four segments shown in gray and labeled 1–4 are non-coding regions with significant similarity between mouse and human genomic sequences and were further studied for potential regulatory function.

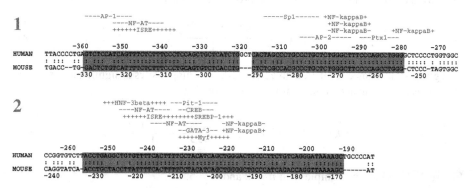

Fig. 2. Putative TF binding sites in the IL28A promoter. The positions are given with respect to the initial ATG.

the use of *phylogenetic footprinting*, a technique that has proven a valuable resource for the identification of functional regulatory elements [2]. However, the results do not identify the significance of TF binding sites in the sense that their combined occurrence is surprising. In order to determine the significance of the binding sites, we apply a statistical model for the clustering of sites within the conserved regions.

2   Prediction of Orthologous Binding Site Clusters: Orthocluster

2.1   Statistical Model

We use a simple statistical model that is based on the assumption that the occurrence of sites for the individual TFs can be modeled by a Poisson distribution [3]. For a genomic region of interest, the significance of the occurrence of sites is determined by comparison with the site frequencies in a background sequence data set. For the individual transcription factors TF_i included in our model, we compute the background rates (site frequencies) λ_i by determining putative binding sites in $10,000$ basepair upstream regions of $7,395$ human genes. The sites are required to be conserved between human and mouse, therefore the background rates are generally lower than if one simply considered the human upstream regions. In this way, the model explicitly includes cross-species comparisons in the computation of statistical significance. To our knowledge, most currently available methods use orthologous sequence alignments only as a post-processing filter that improves the specificity of the algorithm [4, 5] . (Two exceptions are the program Stubb [6] and the ModuleSearcher/ModuleScanner tools available via the TOUCAN workbench [7].)

In our approach, multiple sites for n TFs can be described by a superposition of Poisson distributions [3], with the rate λ for the superposed process given by

$$\lambda = \sum_i^n \lambda_i .\tag{1}$$

Then for a cluster containing k sites, the probability density $p_k(l)$ is given by

$$p_k(l) = \frac{(\lambda l)^{k-1}\lambda e^{-\lambda l}}{(k-1)!} .\tag{2}$$

At a given significance level P, a given cluster containing k sites is significantly shorter than expected by chance alone, if for a window of size L,

$$P(N_L \geq k) = \int_0^L p_k(l)\, dl = 1 - \sum_{j=0}^{k-1} \frac{e^{-\lambda L}(\lambda L)^j}{j!} < P .\tag{3}$$

Figure 3 illustrates an example with $n = 2$ and $k = 5$.

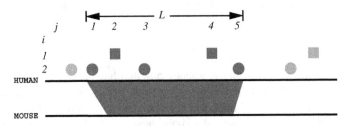

Fig. 3. Example with $n = 2$ and $k = 5$.

Many significance tests are performed when we search a large number of upstream regions for many different combinations of TF binding sites. We do not address this issue here, which would require a correction of the P-values in order to avoid Type I error [3]. Hence our P-values have to be regarded as relative rather than absolute measures.

3 Application of Orthocluster to IL-28A

For $n = 5$, and using a superset of 40 TFs, the most significant cluster identified in the IL28A promoter consists of ISRE (-353), NF-κB (-299), NF-κB (-284), ISRE (-247), and NF-κB (-222), where the positions are given relative to the initial ATG. (For $n < 5$, subsets of this cluster were detected.) Comet [8] and Cluster-Buster [9] were used for comparison and gave consistent results.

4 Application to Large Data Sets

In the previous example, we started with a superset of 40 TFs and we could identify a significant cluster of binding sites for a small set of TFs known to be associated with immune related function (ISRE and NF-κB). In general, however, this approach can be expected to be problematic for assigning genes to functional categories, due to redundancies in the DNA recognition sequences of TFs and the combinatorial nature of the problem.

Selection of subgroups of TFs based on the biological contexts to be studied is likely to decrease the amount of noise in the results. One approach towards this goal is the definition of subsets of TFs associated with expression patterns of interest: In the following we use two disjoint subsets that have been identified in association with hepatocyte-specific gene regulation (C/EBP-α, HNF-1, HNF-3, and HNF-4 [10]) and immune-related function (AP-1, ISRE, NF-κB, and STAT [11]). We assign a gene to one of multiple categories (in our case either liver or immune) depending on which TF model yields the best P-value.

The extraction of human upstream regions from genomic sequence and the identification and extraction of their mouse orthologs have been implemented as automated processes.

First tests were performed using the liver sequence set from [4] and an immune-related set. The algorithm was able to correctly classify the majority of genes known to fall into those categories. Details are given in Appendix B.

Subsequently, the approach was applied to a set of 1815 annotated RefSeq mRNA entries. The results are shown in Table 1.

The 15 most significant results that were classified as liver-specific (immune-related) contain 9 (11) true positive in the sense that they are annotated as liver-specific (immune-related) genes, giving a sensitivity $\geq 60\%$. A more detailed analysis of the results regarding annotated TF binding sites is given in Appendix C.

Tests showed that the sensitivity is largely robust with regard to a variation of the weight matrix score threshold in the range of 65–75%.

The specificity depends on the setting for the P-value threshold. Based on a review of our test data by biologists (data not shown), we used $P \sim 10^{-7}$ as a P-value threshold for our screens of unannotated genes.

Table 1. For each model, the best 15 hits from a set containing 1815 RefSeq entries are shown. True positives [i.e. liver-specific genes in (a) and immune-related genes in (b)] are denoted by +.

(a) Liver Results

Acc. #	P	HUGO symbol	description	+
NM_000384	1.3×10^{-11}	APOB	apolipoprotein B (including Ag(x) antigen)	+
NM_000096	1.6×10^{-11}	CP	ceruloplasmin (ferroxidase)	+
NM_000562	2.1×10^{-11}	C8A	complement component 8, alpha polypeptide	
NM_005807	4.7×10^{-11}	PRG4	proteoglycan 4	+
NM_013371	5.3×10^{-11}	IL19	interleukin 19	
NM_005010	5.8×10^{-11}	NRCAM	neuronal cell adhesion molecule	
NM_000488	1.9×10^{-10}	SERPINC1	serine (or cysteine) proteinase inhibitor	+
NM_000583	4.2×10^{-10}	GC	group-specific component	+
NM_004950	6.4×10^{-10}	DSPG3	dermatan sulfate proteoglycan 3	+
NM_018914	7.6×10^{-10}	PCDHGA11	protocadherin gamma subfamily A, 11	
NM_000477	1.1×10^{-9}	ALB	albumin	+
NM_001962	1.1×10^{-9}	EFNA5	ephrin-A5	
NM_001756	1.3×10^{-9}	SERPINA6	serine (or cysteine) proteinase inhibitor	+
NM_004967	2.0×10^{-9}	IBSP	integrin-binding sialoprotein	
NM_001134	2.1×10^{-9}	AFP	alpha-fetoprotein	+

(b) Immune Results

Acc. #	P	HUGO symbol	description	+
NM_002176	4.5×10^{-12}	IFNB1	interferon, beta 1, fibroblast	+
AF122906	2.4×10^{-11}	IL18BP	interleukin 18 binding protein	+
NM_001503	6.2×10^{-11}	GPLD1	glycosylphosphatidylinos. spec. phosphol. D1	+
NM_002416	3.2×10^{-10}	CXCL9	chemokine (C-X-C motif) ligand 9	+
NM_013243	3.5×10^{-10}	SCG3	secretogranin III	+
NM_000395	5.0×10^{-10}	CSF2RB	colony stimulating factor 2 receptor, beta,	
NM_001561	6.7×10^{-10}	TNFRSF9	tumor necrosis factor receptor superf., m. 9	+
NM_003326	1.4×10^{-9}	TNFSF4	tumor necrosis factor (ligand) superf., m. 4	+
NM_000589	1.5×10^{-9}	IL4	interleukin 4, transcript variant 1	+
NM_000371	1.5×10^{-9}	TTR	transthyretin (prealbumin, amyloidosis type I)	
NM_020525	1.8×10^{-9}	IL22	interleukin 22	+
NM_004407	2.3×10^{-9}	DMP1	dentin matrix acidic phosphoprotein	
NM_000074	2.7×10^{-9}	TNFSF5	tumor necrosis factor (ligand) superf., m. 5	+
NM_016584	3.4×10^{-9}	IL23A	interleukin 23, alpha subunit p19	+
NM_004355	3.5×10^{-9}	CD74	CD74 antigen	

5   Conclusions and Outlook

We demonstrated that the described approach can be used on a genomic scale to provide clues about the function of novel or uncharacterized genes.

There are a number of possible notable variations and extensions of the current model.

Instead of anchoring the extraction of the upstream regions at the locations of annotated transcription start sites, a looser definition of an upstream region, for example based on FirstEF predictions [12], may be considered.

We applied the method only to two clusters, associated with hepatocyte-specific gene regulation and immune-related function, respectively. One can, in principle, try to construct models for a variety of other biological contexts by modifying the set of TFs included in the search (e.g. TFs associated with transcriptional regulation of skeletal muscle, including Mef-2, Myf, SRF, TEF, Sp1, and AP-1 [13]). Or one can attempt to design a modified model, as for example a specialized immune model for genes stimulated by type I interferons by focusing the search on ISRE binding sites. The identification of TFs associated with the biology of interest, however, constitutes a non-trivial step. Contributions from ongoing text mining efforts in the Computer Science community [14] have been helpful and can be expected to further aid the characterization of genes in the future.

Acknowledgments

I would like to thank Patrick J. O'Hara, James L. Holloway, James W. West, and the members of the Bioinformatics Department at ZymoGenetics for many helpful suggestions and critical comments.

References

1. Sheppard, P. et al.: IL-28, IL-29 and their class II cytokine receptor IL-28R. Nat. Immunol. **4** (2003) 63–68
2. Wasserman, W.W., Palumbo, M., Thompson, W., Fickett, J.W., and Lawrence, C.E.: Human-mouse genome comparisons to locate regulatory sites. Nat. Genet. **26** (2000) 225–228
3. Wagner, A.: Genes regulated cooperatively by one or more TFs and their identification in whole eukaryotic genomes. Bioinformatics **15** (1999) 776–784
4. Loots, G.G., Ovcharenko, I., Pachter, L., Dubchak, I., and Rubin, E.M.: rVista for comparative sequence-based discovery of functional transcription factor binding sites. Genome Res. **12** (2002) 832–839
5. Lenhard, B., Sandelin, A., Mendoza, L., Engstrom, P., Jareborg, N., and Wasserman, W.W.: Identification of conserved regulatory elements by comparative genome analysis. J. Biol. **2** (2003) 13
6. Sinha, S., Van Nimwegen, E., and Siggia, E.D.: A probabilistic method to detect regulatory modules. Bioinformatics **19** Suppl. 1 (2003) I292–I301
7. Aerts, S., Van Loo, P., Thijs, G., Moreau, Y., and De Moor, B.: Computational detection of cis-regulatory modules. Bioinformatics **19** Suppl. 2 (2003) II5–II14
8. Frith, M.C., Spouge, J.L., Hansen, U., and Weng, Z.: Statistical significance of clusters of motifs represented by position specific scoring matrices in nucleotide sequences. Nucleic Acids Res. **30** (2002) 3214–3224
9. Frith, M.C., Li, M.C., and Weng, Z.: Cluster-Buster: finding dense clusters of motifs in DNA sequences. Nucleic Acids Res. **31** (2003) 3666–3668
10. Krivan, W. and Wasserman, W.W.: A predictive model for regulatory sequences directing liver-specific transcription. Genome Res. **11** (2001) 1559–1566
11. Liu, R., McEachin, R.C., and States, D.J.: Computationally identifying novel NF-κB-regulated immune genes in the human genome. Genome Res. **13** (2003) 654–661
12. Davuluri, R.V., Grosse, I., and Zhang, M.Q.: Computational identification of promoters and first exons in the human genome. Nat. Genet. **29** (2001) 412–417

13. Wasserman, W.W. and Fickett, J.W.: Identification of regulatory regions which confer muscle-specific gene expression. J. Mol. Biol. **278** (1998) 167–181
14. Light, M., Arens, R., Leontiev, V., Patterson, M., Qiu, X., and Wang, H.: Extracting Transcription Factor Interactions from Medline Abstracts. Posters from the 11th International Conference on Intelligent Systems in Molecular Biology, Brisbane, Australia, (2003)
15. Jareborg, N., Birney, E., and Durbin, R.: Comparative analysis of noncoding regions of 77 orthologous mouse and human gene pairs. Genome Res. **9** (1999) 815–824
16. Batzoglou, S., Pachter, L., Mesirov, J.P., Berger, B., and Lander, E.S.: Human and mouse gene structure: comparative analysis and application to exon prediction. Genome Res. **10** (2000) 950–958
17. Fickett, J.W.: Quantitative discrimination of MEF2 sites. Mol. Cell. Biol. **16** (1996) 437–441
18. Wingender, E. et al.: The TRANSFAC system on gene expression regulation. Nucleic Acids Res. **29** (2001) 281–283
19. Krivan, W.: Searching for transcription factor binding site clusters: how true are true positives? Journal of Bioinformatics and Computational Biology, to be published (2004)

Appendix

A   Methods

The pairwise sequence alignments used for Fig. 1 were performed with DBA [15] as well as with an in-house tool based on WU-BLAST, version 2.0, and a local implementation of GLASS [16]. With the default parameter settings, the different tools gave consistent results. GLASS was used for the alignments performed within Orthocluster.

The search for individual TF binding sites was performed with standard position weight matrices [17] drawn from the TRANSFAC database (version 3.0 [18]) as well as several matrices that were assembled in-house. The set of 40 TFs used was selected with the goal to eliminate some of the redundancies in the original TRANSFAC matrix collection and consists of AP-1, AP-2, ARE, BSAP, CCAAT, CEBP, CREB, CdxA, ER, Egr-1, Elk-1, Freac-3, GATA-3, GR, HNF-1, HNF-3β, HNF-4, HSF1, ISRE, Max, NF-1, NF-AT, NF-κB, Oct-1, PPAR-γ, Pax-6, Pit-1, Ptx1, RORα1, SF-1, SREBP-1, SRF, STATx, Sp1, Staf, TATA, Tbet, YY1, p53, and v-Maf.

B   Tests Using Documented Liver-Specific and Immune-Related Genes

Tests were performed with human upstream regions with mouse orthologs for the seven liver specific genes with RefSeq accession numbers NM_000780, NM_000340, NM_000151, NM_000312, NM_000463, NM_000035, and NM_000207. Using a weight matrix threshold of 65%, four sequences were correctly classified (NM_000780, NM_000340, NM_000151, and NM_000035). At a threshold of 70%, this was the case only for NM_000780, NM_000151, and NM_000035.

Tests were performed with human upstream regions with mouse orthologs for for the six immune-related genes NM_024013, NM_000605, NM_002176, NM_002177, NM_000619, and NM_172138. For both weight matrix thresholds used (65% and 70%), all sequences but NM_002177 were correctly classified.

C   Analysis of the Immune-Related Predictions on the Binding Site Level

The detection of individual sites that themselves are true positives ensures that one does not only detect a significant overall signal, but also that the signal is produced for the right reasons [19].

A detailed analysis of the results summarized in Table 2 shows that typically the overall signal is only partly resulting from the detection of annotated binding sites.

Table 2. The best 10 hits for the immune model from a set containing 1815 RefSeq entries are shown. True positives are denoted by +. Documentation of verified binding sites is given in the last column (PubMed identifiers). See `http://srs6.bionet.nsc.ru/ srs6bin/cgi-bin/wgetz?-e+[TRRDGENES4-AC:A00274]` for IFNB1. (*) The detected region in the IL18BP promoter does not match experimentally verified binding sites. (**) Secretogranin III is regulated by CREB, a pattern similar to AP-1, which is included in the model. (***) Transthyretin is regulated by AP-1.

Acc. #	P	HUGO symbol	+	binding sites (PMID)
NM_002176	4.5×10^{-12}	IFNB1	+	URL
AF122906	2.4×10^{-11}	IL18BP	(*)	12482935
NM_001503	6.2×10^{-11}	GPLD1	+	
NM_002416	3.2×10^{-10}	CXCL9	+	12403783
NM_013243	3.5×10^{-10}	SCG3	+	(**)
NM_000395	5.0×10^{-10}	CSF2RB		
NM_001561	6.7×10^{-10}	TNFRSF9	+	12706838
NM_003326	1.4×10^{-9}	TNFSF4	+	
NM_000589	1.5×10^{-9}	IL4	+	12479817
NM_000371	1.5×10^{-9}	TTR	(***)	1870969

PhyloGibbs: A Gibbs Sampler Incorporating Phylogenetic Information

Rahul Siddharthan[1], Erik van Nimwegen[2], and Eric D. Siggia[1]

[1] Center for Studies in Physics and Biology, The Rockefeller University,
1230 York Avenue, New York, NY 10021, USA
[2] Division of Bioinformatics, Biozentrum, University of Basel,
Klingelbergstrasse 50/70, CH-4056 Basel, Switzerland

Abstract. We present a new Gibbs sampler algorithm with the motivation of finding motifs, representing candidate binding sites for transcription factors, in closely related species. Since much conservation here arises not from the existence of functional sites but simply from the lack of sufficient evolutionary divergence between the species, a conventional Gibbs sampler will fail. We compare the effectiveness against conventional methods on closely-related yeast sequences. Our algorithm is also applicable to single-species or phylogenetically-unrelated sequences, and has further improvements over previous Gibbs samplers, including accounting for correlations in the "background" model, an option to search for "dimers" (pairs of motifs with variable spacing), and a "tracking" strategy that allows us to assess the significance of candidate motifs.

1 Introduction

Gene transcription is regulated by transcription factors, proteins that bind upstream of a gene and typically recognise a short conserved pattern, or "motif", in the DNA. The development of motif-finding algorithms to scan regulatory regions and look for overrepresented motifs is thus of great interest.

For a motif finder to be effective, there must be several copies of a motif to find: it is impossible to detect just one copy of a motif without other prior knowledge, and hard to conclude that two fuzzy copies indicate overrepresentation. To increase the number of copies, one option is to examine genes that are known to be regulated by the same factor. This is not always possible, though often hints can be drawn from microarray experiments. But another option is offered by the increasing number of genomes of closely related species that have appeared in the recent past: we can increase the amount of sequence available by looking at regulatory regions of homologous genes in different closely-related species. For example, sequences of four near relatives of the yeast *S. cerevisiae* (namely, *S. kudriavzevii* [1], *S. bayanus, S. mikatae* [1,2], and *S. paradoxus* [2]) have been published, as well as two more distantly related species, *S. castellii* and *S. kluyveri* [1]. Similarly, in addition to the fruit fly *D. melanogaster*, its close relative *D. pseudoobscura* has been sequenced, and fragments of sequence for various other near Drosophila species exist. In the mammalian world, comparative genomics using the human, chimpanzee,mouse and rat genomes appears promising.

E. Eskin, C. Workman (Eds.): RECOMB 2004 Ws on Regulatory Genomics, LNBI 3318, pp. 30–41, 2005.
© Springer-Verlag Berlin Heidelberg 2005

Attempts have been made, for example in the yeast papers by Cliften et al. [1] and Kellis et al. [2] to find motifs using this extra phylogenetic information, but these have been in the nature of phylogenetic "screens" that concentrate on conserved blocks. Here we evolve a method that accounts for both conserved and non-conserved regions in a transparent and consistent way: this is important because known functional binding sites are not always conserved in other species [3, 4]. Our starting point is the Gibbs sampler. This is a Markov-chain Monte Carlo method [5] to sample a phase space by making a choice at each step from numerous possible moves, weighted by their probabilities; it is more computationally intensive than the usual Metropolis algorithm (where a random move is tried and accepted or rejected), but in problems such as this, converges much faster. In the biological motif-finding context it was introduced by Lawrence et al. [6, 7].

In addition to phylogeny, we make other enhancements to the idea of the Gibbs sampler, the most important of which is a "tracking" mechanism to determine the significance of motifs. This is described in detail later.

2   Scoring with Phylogenetic Conservation

Motifs are often represented by "weight matrices" [8] $w_{\alpha n}$, the probability of finding base α (= A, C, G, T) at site n of the motif (summing over α to 1 for each n.) It is assumed that different columns of the weight matrix are independent.

Most intergenic DNA is probably not functional; non-functional sites are assumed to be described by a "background model" instead. Rather than use raw base counts, we use a background model that incorporates correlations, described below.

With closely related species, much sequence is conserved not because of functionality (presence of binding sites) but because the species are too recently diverged to have mutated significantly. Typical motif finders described above can get misled by such meaningless conservation; we want to account for phylogenetic conservation and adjust the scoring of motifs for this.

When one tries to align these intergenic regions, using alignment tools such as Clustalw [9] or Dialign [10], one finds that there are large blocks of sequence that are highly conserved, interspersed with significant blocks of unconserved, inserted or deleted sequence between different species. We want to treat the non-conserved blocks just as we would an independent sequence, while accounting for phylogeny in the conserved blocks.

We use the following strategy: First, we identify phylogenetically conserved blocks in the sequence, using the alignment tool Dialign [10], with rather stringent parameters for identifying conserved blocks, so that aligned regions are typically rather highly conserved.

Then, we parse the sequence into "windows" – possible sites for motifs, all necessarily all of the same length not counting gaps. In the absence of phylogenetic alignment, "windows" are simply stretches of sequence of length L (the length of the motif), as in figure 1. With phylogenetically aligned regions, windows extend across all aligned sequences – that is, if a base in one sequence has an aligned base in another sequence, that other sequence must be part of the window, as illustrated in figure 2. Windows must be "consistent": there are no "gaps" and pairs of aligned bases are always a consistent distance apart. Thus, we are assuming that a putative binding site in an aligned block is a candidate site either in all the aligned sequences, or in none of them.

```
ac|tggaatagca|tga|tgcgtgcaaa|tgatc
aatactataga|tatcaccaaa|tactatcat
a|tacaacaata|ctgatgaccataacacaaa
```

Fig. 1. Independent sequences (without dialign constraints) and an example of a configuration where four windows have been placed

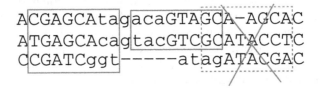

Fig. 2. Aligned sequences in the fasta format output by dialign; only vertically aligned upper-case letters are assumed to have originated from a common ancestor. The dashes are inserted to align the uppercase letters; lowercase letters are not aligned and may be moved through adjacent dashes, for example the "atag" in the last line can be moved before the preceding dashes adjacent to the "ggt", if one wants to place a window at those sites; but the subsequent uppercase letters cannot be moved. Two legitimate windows (solid borders) are shown, one encompassing all three sequences, the other encompassing two of the three. In addition, an illegitimate window is shown (dashed border) – illegitimate because it contains a deletion in a conserved block, which violates our assumption that a motif in a conserved block must be found in all species

The multiple alignment defines in this way the space of possible windows representing binding sites. These sites are sampled uniformly: a window spanning multiple sequences in an aligned block is sampled as often as a single-sequence window. A "configuration" is a particular choice of selected windows representing binding sites.

The "score" of a configuration is the probability that all the windows in that configuration were drawn from the same weight matrix, divided by the probability that all of them were drawn from a background model. (Alternatively, one could use the probability that these windows were sampled from the weight matrix, multiplied by the probability that all sites not in these windows were sampled from the background – this gives the probability of drawing the entire sequence given the current configuration of windows. It is convenient, however, to normalise this by dividing by the probability that the entire sequence was sampled from the background with no weight matrices; this gives our score.) The Gibbs sampler samples for this score; high-scoring configurations represent likely locations for binding sites.

First we describe the score for single-sequence (phylogenetically independent) windows: For a given w, the probability that the windows in a configuration C were all sampled from w is

$$P(C|w) = \prod_{i=1}^{N} \prod_{n=1}^{L} w_{\alpha_{i,n}n} \tag{1}$$

where the i'th motif has base $\alpha_{i,n}$ at position n. Since we don't know the weight matrix, we integrate over the space of all possible weight matrices (that is, over each component $w_{\alpha n}$ with $0 \leq w_{\alpha n} \leq 1$ and $\sum_{\alpha=A,C,G,T} w_{\alpha n} = 1$). This integral can be done exactly:

$$\int_w \prod_\alpha w_\alpha^{n_\alpha} = \frac{3! \prod_\alpha n_\alpha!}{(N+3)!}$$

where $N = \sum n_\alpha$ is the total number of windows and n_α is the 'base count' of base α, that is, the number of windows where base α appears at that position. Alternatively one can include a "prior probability" for weight matrices: $P(C) = \int_w P(C|w)P(w)dw$. With a suitable choice of $P(w)$ [11] this approach is equivalent to that of Liu et al. [7].

The probability that these windows were sampled from a background model is given by eq. (1) with background probabilities b replacing the weight matrix elements w. No integral is required since background probabilities are known.

For multi-sequence aligned windows, we modify the scoring for the window, assuming (irrespective of whether bases in it are uppercase or lowercase) that the bases in it, whether sampled from a weight matrix or from a background model, did not arise independently but evolved from a common ancestor.

We assume a "star" topology, with all species descending from a common ancestor; more general treatments along the same lines are possible, but complicated. Thus, looking at one column, each base in that column is a descendant of an ancestral base a. Assuming mutation rates m_i and assuming divergence a time t ago, the probability that a base in the i'th descendant is unmutated is $e^{-m_i t} = \mu_i$. The probability that it is mutated is $1 - \mu_i$. After mutation, once selection has operated and fixation occurred, the base is again represented by a sample from the same weight matrix (since the protein is generally unchanged in such close relatives). These considerations give us the following expression for the probability $P(W|w)$ that these bases in one column of the window W, descended from a common ancestor, were sampled from the same weight matrix, in terms of a "transition probability" $T(\alpha_i|a)$ that the base α_i evolved from an ancestor a [12]:

$$T(\alpha_i|a, \mu_i) = [\delta_{a\alpha_i}\mu_i + (1 - \mu_i)w_{\alpha_i}] \tag{2}$$

$$P(W|w) = \sum_{a=A,C,G,T} w_a \prod_{i=1}^N T(\alpha_i|a, \mu_i) \tag{3}$$

(position index n omitted for simplicity). The full probability is a product of such factors over all columns. The expression is generally plausible and has the correct limits for $\mu \to 0$ (infinitely diverged species, i.e. independent sequences) and for $\mu \to 1$ (zero divergence, i.e. identical species). Note also that the transition matrix $T(\alpha_i|a)$ has the correct multiplicative property if one inserts an intermediate unknown ancestor: $\sum_b T(\alpha_i|b, \mu_1)T(b|a, \mu_2) = T(\alpha_i|a, \mu_1\mu_2)$. For further discussion, see reference [12].

For a set of windows, the probability that all these windows were sampled by the same matrix is given by a product of factors, one for each window, monomial as in eq. (1) for single-sequence windows and polynomial as in eq. (3) for multi-sequence windows. The total expression is thus likely to be a complicated polynomial, and is then integrated over all weight matrices w. (In practice, we use approximations for this integral, which we have verified are accurate.) Again, for the background scores, one substitutes w with background probabilities b and does not integrate.

A word about background probabilities: one commonly uses the raw "base counts" for each base. These are $1/4$ each in the simplest assumption, but A and T are more frequent in practice than C and G. We found it beneficial to instead use conditional probabilities for the occurrence of each base with the n preceding bases (in other words, assume a Markov model for the sequence.) That is, to calculate the background probability of the C in the sequence AGC, with two-neighbour correlation, we use

$$P(\text{C}|\text{AG}) = \frac{N(\text{AGC})}{N(\text{AGA}) + N(\text{AGC}) + N(\text{AGG}) + N(\text{AGT})} \tag{4}$$

where $N(\text{AGC})$ is the actual number of occurrences of the string AGC in the sequence, etc. To calculate these numbers, it is preferable to use a larger dataset than the sequence of interest: for example, all sequenced intergenic regions in the organism, if available. If not enough sequence exists, a pseudocount of raw base counts may be added.

3 Implementation of the Sampler

With the above framework for scoring configurations of windows, we can implement the Gibbs sampler in a straightforward manner. We typically sample for multiple motifs at a time, which in effect means assigning different "colours" to the selected windows; each colour is scored separately and the total score is a product over all colours.

We start from a random configuration. At any instant in time the configuration is a set of selected windows, in different colours. The moveset takes us from one configuration to another, possibly changing the number of windows, the number of colours, or both, and is designed to satisfy "detailed balance" so that, in the long time limit, each configuration would be visited a fraction of the time proportional to its score.

The movesets we developed can follow several different strategies, with two key kinds of moves involved, which we call the "window move" and the "colour move":

1. We can restrict the total number of colours, and the total number of windows, rigidly. At each step we will "move" a window and optionally recolour it to one of the other existing colours: that is, we will pick a window at random, remove it, and sample from all new places where it can be placed, and all possible colours it can have at the new location. We call this the "window move". To preserve the number of colours while maintaining detailed balance, we require that the window being moved was not the only one in its colour; if not, we do nothing but increment the counter. This conserves total colour number and total window number, but not the number of windows per colour. It is also a highly optimal move in escaping local minima by "freeing" windows that are blocked by differently-coloured, suboptimally-placed windows.

2. We can change the number of windows, and the number of colours. To do this, we alternate the "window moves" with another kind of move which we call the "colour move". Here, we pick a window; if it is "blocked" (an overlapping window is coloured), we make no move, but increment the time counter (this is necessary

to preserve detailed balance); otherwise, sample from all possible colours it can be given, which may be the "null" or "background" colour (ie removing a window), one of the existing colours, or a new colour (if the window was already the only one in its colour, "new colour" means the same colour as earlier). This move conserves neither colour number nor window number, but preserves detailed balance. By itself, it does not do a good job of detecting motifs (though it ought to do that in the infinite-time limit, since it is ergodic and satisfies detailed balance); rather, its role is to expand the "colour space" and "window number space" of the system, while the "window moves" do the job of actually aligning motifs together.

This strategy of mixed moves will tend to populate the sequence with as many windows and as many colours as necessary to maximise the score, simply for entropic reasons (there are more states with a lot of windows than with a few windows). Thus, each configuration will be a "parse" of the sequence into similar motifs. A benefit is that one doesn't need to guess how many copies of a motif one is looking for: if good copies exist, they will be added to existing colours, otherwise new colours will be created. The disadvantages are that this strategy may "split" fuzzy motifs into several smaller groups, and the number of motifs it yields may be so large as to be unwieldy.

3. We can use the "colour moves" to allow a flexible colour/window number, but use a "chemical potential" to constrain the number of windows. Thus, every extra added window has a cost. Alternatively, we can introduce an entropic "correction factor" to take account of the fact that there are more states with $n + 1$ windows than with n windows (until n becomes large); the factor is easy to calculate for a single sequences, harder for multiple unrelated sequences and very hard for the phylogenetically aligned case. We can also place a rigid upper limit on the total number of colours.

To use the first strategy, we need initial values for N, C and the expected motif length L. It is fine to somewhat overestimate C and L, and advisable to somewhat underestimate N compared to the number of binding sites one actually expects to see. Typically one has one or two strong motifs, and the redundant colours while sampling then act as "buffers" to control the number of motifs actually picked by the sampler.

Regardless of strategy, in addition to the above moves, there are two other moves that we use to improve performance. A "global shift" move samples all possible shifts of an entire colour by a fixed amount; this is necessary because once the program finds a good placement of windows that is shifted relative to its optimal position, it is impossible to correct this by single-window shifts: the program will stay stuck in a local minimum. A "maskbit flip" move turns on sampling of mask bits for columns that indicate whether that column is to be scored or not; this is inadvisable for short motifs but is particularly useful for long motifs that have intervening fuzzy regions (such as occur in bacterial sequences), where it would be preferable not to score the fuzzy columns.

4   An "Anneal and Track" Strategy for Assessing Significance

In sampling for a long time, all configurations will be visited in the infinite-time limit with a frequency proportional to their score. To find the configuration with the best score, we can anneal (that is, raise the score to a power β representing a fictitious inverse temperature, and slowly raise β). If there is one overwhelmingly good configuration, a well-implemented anneal will typically find it easily; if there are several comparably good configurations (for example, more binding sites than one is searching for, or different motifs corresponding to different transcription factors), the anneal will randomly choose one of them.

Annealing in the framework of our moveset gives us a good candidate set of windows (binding sites); and we may also have candidate windows by comparing multiple anneals, or from other sources altogether; but it is desirable to assess their significance in some way. The approach below does this, and as a side benefit, also assigns significance to other sites which may not have been in our candidate set.

To do this, we set up a "labelled list" of the windows that we want to track. We could even up several such "labelled lists", with different labels A, B, C..., and track them simultaneously. Then we sample for a long time without annealing ($\beta = 1$), doing the following:

1. At each time step, there is a set of colours each with a set of selected windows. For each labelled list A, we associate *one* of the current colours with that labelled list. Unless we are exceptionally lucky, none of the current colours will precisely match the labelled list A: there will be windows in that colour that are not in the list, or vice versa. So

    (a) We examine each colour for windows from our labelled list, with all possible consistent shifts and orientations;

    (b) For each colour and shift, we note the windows from the labelled list that appear in that colour with that shift (all these windows must have the *same* shift, or opposite shift if the orientation is opposite);

    (c) We calculate a total "importance score" of these windows to that colour by totalling the "cost" of removing each of these windows from the colour (that is, the ratio of the score of the colour with that window, to the score of the colour without that window);

    (d) Finally, we choose the colour and shift that gained the highest "importance score" by the above definition. The chances of this importance score being degenerate are negligible, but if it happens, we can make an arbitrary choice.

    This defines, at every instant, for each labelled list A, a unique colour associated with it, which we call $C(A)$, and an associated global shift, $S(A)$. This is the best match we have to our labelled list in the current configuration.

2. For every window w in the sequences being sampled, whether in the labelled lists or not, we maintain a set of counters, one for each labelled list, $N(w, A)$. This is an $N_w \times N_l$ matrix, where N_w is the number of windows and N_l the number of labels.

3. At each time step, for each label A, we go through the windows in the corresponding colour $C(A)$, shift each window w by $-S(A)$ to align properly with the labelled list if $S(A) \neq 0$, call the shifted window w', and increment the corresponding counter $N(w', A)$.

Finally, we divide each counter by the total number of timesteps; this gives, for each window w and each label A, a time average of the function

$$f(A, w) = 1 \text{ if } w \in C(A)$$
$$= 0 \text{ otherwise}$$

which measures how often the window w was "co-clustered" with the labelled list A. In the infinite-time limit, this is (since each state is visited a fraction of time proportional to its score)

$$\frac{T(\text{co-clustered})}{T(\text{total})} = \frac{\sum_{S \text{ where w co-clustered with A}} P(S)}{\sum_{\text{all } S} P(S)} \tag{5}$$

which is intuitively the probability that w was sampled from the same weight-matrix as the windows in the labelled list A.

Thus, for each label A, on sorting the corresponding row of $N(w, A)$ we get a list of windows ordered by the probability that they "belong" with the list A.

This turns out to be a very useful strategy, not only in finding unknown motifs, but in assessing known ones: for example, one can "seed" a sequence with one or two short sequences of known motifs, and track the known motifs to see who else gets clustered with them.

5   Performance of the Sampler

We have tested the sampler both on synthetic data generated according to the model we assume, and on actual genomic data from the five closely related yeast species *S. cerevisiae, S. paradoxus, S. bayanus, S. mikatae, S. kudriavzveii* using genes with known motifs.

The "tracking" mechanism described earlier, apart from being a useful tool in practice for significance estimates of found motifs, is also a useful benchmarking tool when looking for known motifs: tracking numbers are a measure, in a quantitative and directly relevant way, of the probability for each tracked motif site that it was drawn from the same weight matrix as the others.

On synthetic data, for purposes of benchmarking, we track the known positions of the weight matrices. In other words, we sample for a long time, and collect statistics on what fraction of the time the known motif sites actually hang together. (Thus, we are not benchmarking the anneal, which is the motif-finding step; however, sites that hang together on sampling are always found in an anneal, though the reverse is not true.)

All tests were with five sequences, generated from a single ancestral sequence of length 500bp and a certain number n of embedded copies of a motif described by a weight matrix; the polarisation of the weight matrix, the number n of embedded copies, and the phylogenetic conservation probability μ were varied.

Phylogibbs results are clearly and consistently better than results with the sampler without considering phylogeny (treating the sequences as independent), except when the phylogenetic conservation probability μ is so high that it approaches the weight matrix polarisation. Results are in figures 3 and 4.

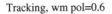

Tracking, wm pol=0.6

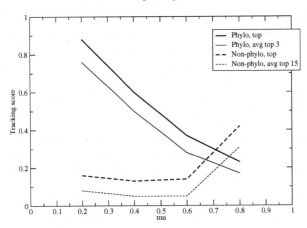

Tracking, wm pol = 0.7

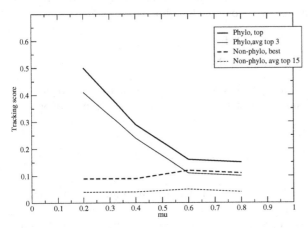

Tracking, wm pol=0.8

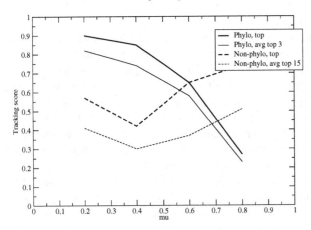

Fig. 3. Plots showing the fraction of time the Gibbs sampler clusters known motifs together, for five sequences of length 500 each descended from an ancestor, with five embedded copies of a motif represented by a weight matrix of polarisation as indicated, mutated according to our model with conservation probability μ. The known positions of the embedded weight matrices were tracked. Shown are the tracking scores of the best motif ("top") and the best 3 (phylo) or 15 (non-phylo) motifs ("avg"), each averaged over three runs, as a function of μ. Except when μ is high enough (0.8) to compare with or exceed the polarisation, phylogibbs performs clearly better. In particular, even when one or a few motifs may get comparable tracking scores in the absence of phylogeny, the average tracking score (here averaged for the top 3 motifs with phylogeny, or $3 \times 5 = 15$ motifs without) is far better with phylogeny. (The top 3, or 15, were averaged rather than all 5, or 25, because in many cases the tracking scores of the remainder in the non-phylogibbs case fell below the reporting threshold for the program)

Tracking scores, mu=0.5, wm pol=0.75

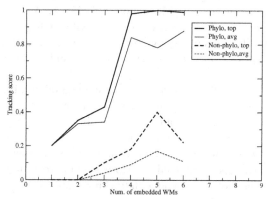

Fig. 4. Tracking scores, with and without phylogeny, for synthetic data with $\mu = 0.5$, and n weight matrices embedded, with wm polarisation 0.75. Shown are the best tracking score ("top"), and the tracking score averaged over all motifs ("avg"), as a function of n.

On real data, this is usually the case, too; sometimes, when motifs are very numerous and prominent, both versions perform very well, and occasionally phylogibbs does worse, apparently because known motifs lie in conserved regions but are mutated in other species. We studied a few genes from *S. cerevisiae*, with well-documented regulatory sites, that we were interested in for other reasons. We assume a uniform conservation probability $\mu = 0.5$ (varying μ from around 0.3 to 0.7 does not make a big difference to results; $\mu = 0.3$ is a reasonable estimate from synonymous substitution in coding regions, but in practice *cis*-regulatory regions seem to have a somewhat higher conservation rate.) The anneal stage searched for 4 different colours, and 16 possible regulatory sites (avg: 4 per colour), in the phylogibbs case. Because on average 70%–80% of the *cis*-regulatory region seems to fall in aligned blocks, in the non-phylogibbs case we searched for 64 possible sites (avg: 16 per colour). The results are as follows:

- CLN3 (YAL040C): There are four well-defined copies of an element that has been called the "daughter delay element" [13], which has been implicated in the delay in budding in daughter cells. This element has consensus CCWYWGCATTTC and is instantly picked up by phylogibbs with tracking score 1.00. However, without incorporating phylogeny this motif is often not picked up at all in the initial anneal, and when it is picked up it gets a lower tracking score. Apparently this is because there are several copies of similar motifs such as CCWWW... (half of the MCM1 dimer site, which appears in several places upstream of this gene), CCNNNGC, and SSATTTC, some of which have neighbouring sequence similar to the DDE, and these tend to lead the sampler astray.
- HO (YDL227C): We used the first 1000 bp upstream region of this gene, though it has one of the longest *cis*-regulatory regions in cerevisiae. With or without phylogeny we retrieve numerous copies of the SBF binding site [15] (consensus: CAC-GAAA) with tracking score 1.00 for numerous copies; the MATα2 site TTACATCA is also retrieved with tracking score 1.00 with phylogeny, but without phylogeny our runs did not retrieve this motif.

– CLB1 (YGR108W): This gene is known [14] to be regulated by Ndt80 and contains the middle sporulation element (MSE) motif GWCACAAA in its *cis*-regulatory region, but not very strongly. We recover the motif with phylogibbs, but with tracking scores of 0.40-0.50. Without phylogibbs, the anneal yields ambiguous results (the motif is mixed up with other sites) and the tracking yields nothing above the reporting threshold (0.05).

– NDT80 (YHR124W): This is a key gene in sporulation [14, 16, 17] and contains the MSE in its *cis*-regulatory region; the MSE is bound by the Ndt80 gene product itself, as well as by Sum1. We find the motif with or without phylogeny, but in this case the performance is better without phylogeny: tracking scores are around 0.99 without, or 0.6 with. (However, if we use phylogibbs but lower μ to zero – that is, independent sequences but aligned – performance improves to perfect levels: that is, all copies of the motif are found with tracking score 1.00.)

The NDT80 example brings up one worthwhile point: the improvement in performance of phylogibbs comes partly from improved scoring of phylogenetically related sequences, but a significant part of the improvement is merely the much smaller state space when one aligns sequences as we do before sampling. There is a significant reduction in entropy: many configurations that are not likely positions of binding sites are simply removed from the state space. (One can, of course, contrive examples where the reduced state space hurts rather than helps because important configurations are being removed, for example because most motifs occur in conserved blocks but are mutated in all species but one; it's unlikely this is a common problem in practice.)

Availability of the Code

The code is available for download on
`http://www.physics.rockefeller.edu/~siggia/software/phylogibbs/`

Support

This work was supported by NSF grant DMR 0129848.

References

1. Cliften, P., Sudarsanam, P., Desikan, A., Fulton, L., Fulton, B., Majors J., Waterston R., Cohen B. A., Johnston M. Science **301** (2003) 71–6.
2. Kellis M. Patterson N., Endrizzi M., Birren B., Lander E. S. Nature **423** (2003) 241–54.
3. Dermitzakis E. T., Bergman C. M., Clark A. G. Mol Biol Evol. **20**(5) (2003) 703–14.
4. Emberly E., Rajewsky N., Siggia E. D. BMC Bioinformatics **4**(1)(2003) 57.
5. Liu J. S.: *Monte Carlo Strategies in Scientific Computing*. Springer-Verlag (2001).
6. Lawrence C. E. , Altschul S. F., Bogouski S. F., Liu J. S. , Neuwald, A. F. , Wooten, J. C. Science **262** (1993) 208–214.
7. Liu J. S., Neuwald A. F., Lawrence C. E. J. Amer. Stat. Assoc. **90** (1995), 1156–1170.
8. Durbin R., Eddy S., Krogh G., Mitchison G. *Biological Sequence Analysis*, Cambridge University Press (1998).
9. Thompson J. D., Higgins D. G., Gibson T. J. Nucleic Acids Res. **22** (1994) 4673–4680.

10. Morgenstern, B. Bioinformatics **15** (1999), 211–218.
11. van Nimwegen E., Zavolan M., Rajewsky N., Siggia E. D. Proc. Nat. Acad. Sci. **99** (2001) 7323–7328.
12. Sinha S., van Nimwegen E., Siggia E. D., Proceedings of the 11th international conference on Intelligent Systems for Molecular Biology (2003).
13. Laabs T. L., Markwardt D. D, Slattery M. G, Newcomb L. L., Stillman D. J., Heideman W. Proc Natl Acad Sci USA. **100**(18) (2003) 10275–80.
14. Chu S., Herskowitz I. Mol Cell **1** (1998), 685–696.
15. Breeden L., Nasmyth K. Cell **48** (1987) 389–97.
16. Hepworth S. R., Friesen H., Segall J. Mol Cell Biol. **18** (1998) 5750–61.
17. Chu S., DeRisi J., Eisen M., Mulholland J., Botstein D., Brown P. O., Herskowitz I. Science **282** (1998) 1421.

Application of Kernel Method
to Reveal Subtypes of TF Binding Motifs
Causal Analysis of Gene Expression Data

Alexander Kel[1,3], Yury Tikunov[2], Nico Voss[1],
Jürgen Borlak[4], and Edgar Wingender[1,5]

[1] BIOBASE GmbH, Halchtersche Str. 33, D-38304 Wolfenbüttel, Germany
{ake,nvo,ewi}@biobase.de
http://www.biobase.de
[2] Institute of Cytology and Genetics SB RAN, Lavrentyev pr., 10, 630090, Novosibirsk, Russia
kel@bionet.nsc.ru
[3] United Institute of Geology,Geophysics and Mineralogy SB RAN, Koptyug pr., 3,
630090, Novosibirsk, Russia
tikunov@uiggm.nsc.ru
[4] Fraunhofer Institute (Fh-ITEM) of Toxicology and Experimental Medicine,
Center for Drug Research and Medical Biotechnology,
Nikolai-Fuchs-Str. 1, D-30625 Hannover, Germany
borlak@item.fraunhofer.de
[5] Dept. of Bioinformatics UKG, University of Goettingen,
Goldschmidtstr. 1, D-37077 Goettingen, Germany
e.wingender@med.uni-goettingen.de

Abstract. Transcription factor binding sites often contain several subtypes of sequences that follow not just one but several different patterns. We developed a novel sensitive method based on kernel estimations that is able to reveal subtypes of TF binding sites. The developed method produces patterns in form of positional weight matrices for the individual subtypes and has been tested on simulated data and compared with several other methods of pattern discovery (Gibbs sampling, MEME, CONSENSUS, MULTIPROFILER and PROJECTION). The kernel method showed the best performance in terms of how close the revealed weight matrices are to the original ones. We applied the Kernel method to several TFs including nuclear receptors and ligand-activated transcription factors AhR. The revealed patterns were applied to analyze gene expression data. In promoters of differentially expressed genes we found specific combinations of different types of TF binding patterns that correlate with the level of up or down regulation.

1 Introduction

Sites in genomic DNA that serve as targets for binding of a certain transcription factor (TF) share common patterns that are often described by consensus sequences or position weight matrices (PWMs). Elucidation of the structure of TF binding sites is a very important problem because it enables to understand the mechanism of gene regulation. Several methods have been developed in the recent years for identification of patterns shared by a set of functionally related sequences, e.g. CONSENSUS

E. Eskin, C. Workman (Eds.): RECOMB 2004 Ws on Regulatory Genomics, LNBI 3318, pp. 42–51, 2005.

(Hertz and Stormo, 1999), Gibbs Sampler (Lawrence et. al., 1993), MEME (Bailey and Elkan, 1994), ANN-Spec (Workman and Stormo, 2000), PROJECTION (Buhler and Tompa, 2002), combinatorial approaches (Pevzner and Sze, 2000), MULTI-PROFILER (Keich & Pevzner, 2002). A combination of these methods was used for identifying target sites of cooperatively binding factors (Thakurta and Stormo, 2001). The methods that are able to reveal patterns in the form of PWMs are of the most interest now. Position weight matrices currently are the state of the art in modeling the structure of TF binding sites. They are clearly superior to the consensus description. More complex models such as HMM showed quite good performance (Ellrott et al., 2003), but their application is limited now to a few TFs with the high number of known sites.

It is known now that often one set of TF binding sites may contain several subtypes of generally similar but different patterns. This happened in most cases because of the lack of knowledge. Often we don't know what particular isoform of the factor, which homo- or heterodimer, modified variant, or in a complex with which co-factor or other cooperating factor, it actually binds to *in vivo*. Methods for revealing such patterns by sub-clustering of the sets of sequences are urgently needed.

We have developed a novel method for discovery of subtypes of patterns based on the kernel estimation of a probability density function. A first variant of this method was used for the investigation of aligned sequences near the start of transcription of eukaryotic genes (Tikunov and Kel, 2000). Here we present the improvement of this method. It can now be applied to the analysis of unaligned nucleotide sequences. Using simulation of random sequences with implanted sites we have compared the developed kernel method with several other methods such as Gibbs sampling (http://bayesweb.wadsworth.org/gibbs/) (GIBBS), MEME (http://meme.sdsc.edu/meme/), CONSENSUS (http://ural.wustl.edu/), MULTIPROFILER (http://www.cs.ucsd.edu/groups/bioinformatics/) and PROJECTION (http://www.cs.wustl.edu/~jbuhler/projection.html). The Kernel methods showed the best performance in terms of how close the revealed patterns are to the original ones. The Kernel method was able to distinguish two very similar patterns (with more then 30% of the same nucleotides in the consensus) whereas most of the other programs had rather high level of identification errors. The program is available for on-line use at: www.biobase.de/cgi-bin/biobase/cbs2/bin/template.cgi?template=cbscall.html. Source code is available upon request.

The developed method of pattern discovery makes a new advance in our capabilities in interpretation of gene expression data. Regulation of gene expression is accomplished through binding of multiple transcription factors to large regulatory regions of genes. Some of these TFs are specific for a particular tissue, a definite stage of development, or a given extracellular signal, but most transcription factors are involved in gene regulation under a rather wide spectrum of cellular conditions. It is clear by now that combinations of transcription factors rather than single transcription factors drive gene transcription and define its specificity.

We apply the developed Kernel method on the analysis of a set of gene expression data from toxicogenomics studies. We found two pattern subtypes of binding sites for AhR transcription factor, which plays a key role in regulation of genes during antitoxic response of cells. In promoters of AhR activated or repressed genes we found specific combinations of different types of TF binding sites including two AhR pattern subtypes that correlate with the level of up- or down-regulation.

2 Data and Methods

We use two databases in the analysis: TRANSFAC® (BIOBASE GmbH, Wolfenbüttel, Germany) is a database that collects information about gene regulation in eukaryotes based on binding of transcription factors to their target sites; and TRANSCompel® which contains known composite regulatory elements in mammalian genes. We used TRANSFAC® Professional rel.6.4 and TRANSCompel® rel.6.4.

2.1 Kernel Model for Nucleotide Sequences

We consider some set of nucleotide sequences s with length m. Let us denote a whole set of possible sequences as Ω. So the number of elements of set Ω is 4^m. The expected frequency of every possible sequence s is defined with appropriate probability p_s. Let us assign to every sequence some nonnegative weight w_s. We call this function $w(s)=w_s$ the weight kernel. We define for the weight kernel $w(s)$ the averaged weight sum $S_0(w)$ and volume $V_h(w)$, where

$$S_0(w) = \sum_{s \in \Omega} w_s \cdot p_s \tag{1a}$$

$$V_h(w) = \left(\sum_{s \in \Omega} w_s^{1+h} \right)^{1/(1+h)} \tag{1b}$$

$$\Phi_{0,h}(w) = \frac{S_0(w)}{V_h(w)} \tag{1c}$$

The ratio of averaged weight sum $S_0(w)$ to volume $V_h(w)$ is a functional of averaged density $\Phi_{0,h}(w)$ with respect to kernel $w(s)$.

Since $w_s \geq 0$ and $p_s \geq 0$ the Hölder's inequality is true

$$\sum_{s \in \Omega} w_s \cdot p_s \leq \left(\sum_{s \in \Omega} w_s^{y} \right)^{1/y} \cdot \left(\sum_{s \in \Omega} p_s^{z} \right)^{1/z} \tag{2}$$

where $y>1$, $1/y+1/z = 1$. If we put $y = h+1$, $z = (h+1)/h$ then we get from relation (2) to the inequality:

$$\frac{\sum_{s \in \Omega} w_s \cdot p_s}{\left(\sum_{s \in \Omega} w_s^{h+1} \right)^{1/(h+1)}} \leq \left(\sum_{s \in \Omega} p_s^{\frac{h+1}{h}} \right)^{h/(h+1)} \tag{3}$$

The left part of inequality (3) is the averaged density functional $\Phi_{0,h}(w)$. The inequality (2) turns into equality when $w_s^{y} = (c \cdot p_s)^{z}$, where c is an arbitrary positive normalization factor. Hence

$$w_s^{h} = c \cdot p_s \tag{4}$$

It follows that the kernel is the densest when weight kernel function satisfies the equation (4).

If we have some sample of sequences $\Omega_n = \{s_1, s_2, \ldots, s_n\}$ we can construct empirical averaged weight sum $S_n(w)$ and empirical averaged density functional $\Phi_{n,h}(w)$ with respect to kernel $w(s)$ as

$$S_n(w) = \frac{1}{n} \cdot \sum_{s \in \Omega_n} w_s \tag{5a}$$

$$\Phi_{n,h}(w) = \frac{S_n(w)}{V_h(w)} \tag{5b}$$

The mathematical expectation of empirical averaged weight sum $S_n(w)$ is defined with expression (1a) and the expectation of $\Phi_{n,h}(w)$ is defined with expression (1c). In accordance with the law of large numbers the value of functional $\Phi_{n,h}(w)$ converges to the true value in probability under $n \to \infty$. So we can estimate the functional $\Phi_{n,h}(w)$ with any accuracy under $n \to \infty$. If we have some set of kernels $w_\alpha(s) : \alpha \in \Lambda$ the densest kernel defines the probabilities of sequences $p(s)$ in accordance with relation (4). The proposed functional of averaged density $\Phi_{n,h}(w)$ is remarkable because it allows us to reconstruct the probabilities $p(s)$ by putting more weights to the more frequent sequences s. Smoothing parameter h regulates the weights for sequences with different expected probabilities.

Under $h = 0$ the functional $\Phi_{n,h}(w)$ is similar to Parzen-Rosenblatt kernel estimation of probability density which is often used in mathematical statistics for reconstruction of probability density (Rosenblatt, 1956; Parzen, 1962). A.I.Orlov (Orlov, 1991) adopted Parzen-Rosenblatt estimation for analysis of nonnumeric data. However, using of proposed functional $\Phi_{n,h}$ enables the search of the best probability function $p(s)$ suitable not for all space Ω but only for some of its local compact part.

For patterns that are represented by weight matrices $\|f_{jl}\|$ we can apply the described theory and construct an algorithm that allows us to reveal all the patterns in a set of sequences by searching for the clusters which are characterized with local maxima of $\Phi_{n,h}$. We assume that in each such cluster the probability distribution of sequences s is described with this matrix in accordance with independent distribution of nucleotides in different positions $f_s = \prod_{j=1}^{m} f_{jl_j^s}$ where f_s is the frequency of sequence s; f_{jl} is the frequency of letter l in position j (the elements of weight matrix $\|f_{jl}\|$); l_j^s is the letter of sequence s in position j. A consensus s_c that corresponds to a weight matrix $\|f_{jl}\|$ is a sequence that contains the most probable letters in every position. In accordance with equation (2) we define the weight kernel $w_s = (c \cdot f_s)^{1/h}$. Let us put the normalization factor c equal to $1/f_c$, where f_c is the frequency of the consensus sequence. So the weight of consensus sequence equals 1. Hence, the weight kernel may be defined as $w_s = \prod_{j=1}^{m} w_{jl_j^s}$ and $w_{jl} = \left(\dfrac{f_{jl}}{f_{jl_j^c}} \right)^{1/h}$; here, l_j^c is the letter of consensus sequence in position j; w_{jl} is the weight kernel coefficient. If we put $R_s = \ln(f_c/f_s)$ then $R_s = \sum_{j=1}^{m} \gamma_{jl_j^s}$ and $\gamma_{jl} = \ln(f_{jl_j^c} / f_{jl})$. R_s may be considered as a distance of sequence s to consensus and γ_{jl} are the distance coefficients. The larger the distance the less weight it is assigned to this sequence in the model of the given consensus. Smoothing parameter h regulates the dependence of sequence weights from the distance.

Using the inputted above denotations one can derive an equation for the volume of kernel $w(s)$ based on a weight matrix $\|f_{jl}\|$

$$V_h(w) = \left(\sum_{s\in\Omega} w_s^{1+h} \right)^{1/(1+h)} = \left(\prod_{j=1}^{m} \sum_{l\in A} e^{-\gamma_{jl}\cdot(1+h)/h} \right)^{1/(1+h)} \tag{6}$$

where l is a letter of alphabet A. When functional $\Phi_{n,h}$ reaches its maximum its derivative equals zero. So from equation (5a), (5b), (6) by substituting the appropriate denotations we get

$$d\left(\ln\Phi_{n,h}\right) = \frac{1}{h}\cdot\sum_{j=1}^{m}\left(\frac{S_{n,jl}}{S_n} - \frac{e^{-\gamma_{jl}(h+1)/h}}{\sum_{\lambda\in A} e^{-\gamma_{j\lambda}\cdot(h+1)/h}} \right) d\gamma_{jl} = 0 \tag{7}$$

where

$$S_{n,jl} = \frac{1}{n}\cdot\sum_{s\in\Omega_n(jl)} w_s = \frac{1}{n}\cdot\sum_{s\in\Omega_n(jl)} e^{-R_s/h} \tag{8}$$

$\Omega_n(jl)$ is a subsample of sample Ω_n those that the letter l is situated in position j. The $d(\Phi_{n,h})$ equals zero when every member of the sum from right part of equation (7) equals zero. We get from the equation (7) to the following equation system

$$f_{jl} = g \cdot \frac{S_{n,jl}}{e^{-\gamma_{jl}/h}}; \tag{9}$$

The value of g is calculated from the normalization requirement $\sum_{\lambda\in A} f_{j\lambda} = 1$. The densest kernel corresponds to the probabilities $p(s)$ that in the best way describe the distribution of sequences near the local maxima.

2.2 Algorithm for Detection of Multiple Patterns in the Unaligned Sequence Sets

In the algorithm each weight matrix is calculated on the basis of subsequences (words) of length m picked up from the sample (one subsequence from each sequence). The present algorithm is initialized by a random choice of a starting subsequence of the length m from one random sequence of the sample. From all other sequences, one subsequence of the length m is picked up which is the closest to the starting subsequence. On the basis of all these subsequences the initial weight matrix is calculated by just counting of the letters in the appropriate positions. When the initial weight matrix $\|f_{jl}\|$ is built the appropriate weight kernel coefficients and distance coefficients are calculated. The algorithm makes several recursive iterations calculating the next weight matrix coefficients from the equation (9) using currant weight kernel and distance coefficients. The iterations are stopped when no further changes in the matrix elements f_{jl} are observed.

3   Results

First, we tested the Kernel method on a set of simulated data and compared its performance with other methods of pattern discovery. After that we applied it to analyze three sets of TF binding sites. And finally, we use the found patterns in the analysis of gene expression data.

3.1   Comparison of the Kernel Method
with Other Motif Search Algorithms Using Simulated Data

To compare the developed algorithm with other known algorithms we have prepared several samples of simulated data using a setup similar (Workman and Stormo PSB 2000). We generated sets of random sequences in which we implanted sites using predefined weight matrices.

We generated sets of 200 random sequences of the length 24bp each (n=200). Sites of length 10 (m=10) were implanted in a randomly chosen position into half of the sequences (one site in a sequence). The other half of the sequences remains just random. The weight matrix X that was used for generation of the implanted sites contains in every position one nucleotide with the maximal weight ξ varying from 0.65 to 0.95. We call this nucleotide as "consensus" nucleotide. All other nucleotides have got weights $(1-\xi)/3$. In this way we can simulate more conserved or less conserved matrices. Six programs have been compared: the kernel method developed in this work (Kernel), Gibbs sampling program (GIBBS), MEME, CONSENSUS, MULTIPROFILER and PROJECTION.

The default parameters of the programs were used to perform the test. Each program runs several times on different sets of generated sequences. After each run the matrix Y calculated by the program was compared with the original matrix X (distance between matrices is measured by $D = \sum_{jl} (p_{jl}^{initial} - p_{jl}^{calculated})^2$). In order to align matrices we slide matrix Y along the matrix X by 3 positions left and right to find the best fit. (In the case of mismatching we set frequencies 0.25 to shifted part of the X matrix). In Figure 1 we present the results of the comparison of the first four programs. It is obvious that the lower the parameter ξ the more difficult for a program to reveal correctly the matrix.

Programs MULTIPROFILER and PROJECTION approach the precision of the Kernel method, but only for the smallest values of ξ = 0.65 and 0.7 (D = x and y, resp.).

3.2   Application of the Kernel Method to the Sets of TF Binding Sites

AP-1 and CREB. We take a mixture of binding sites for AP-1 and CREB transcription factors from database TRANSFAC. It is known that often transcription factors of

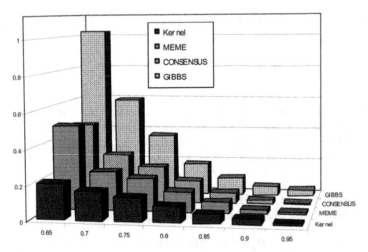

Fig. 1. Result of comparison of four different pattern discovery programs on the sets of simulated sequences with implanted TF binding sites for one matrix; y-axis: the averaged sum of squared differences (D) between revealed matrix and the original one (minimum 5 runs); x-axis: ξ values, that are the probabilities of "consensus nucleotide" in each position of the matrix. The smaller the value the better is the recognition ability of the program.

these two different families bind to the same sites. The analyzed sample contained 155 sequences. Every sequence contained a TF binding site in the center and additional 10 nucleotide flanking both sides of the site.

We have applied our program and have revealed two different patterns. The analysis showed that the overwhelming majority of sequences contain a pattern of length 7 that corresponds to the consensus: "**TGAGTCA**". The second pattern has the length 8 and corresponds to the consensus: "**TGACGTCA**", which differs from the first one by insertion of letter "**C**" in the forth position. It is important to mention that exactly these two patterns correspond to the known consensi of AP-1 site (the first pattern) and CREB site (second pattern). Classification of the investigated sites shows that some of them contain both of these motifs located at different locations but close to each other.

We have applied two other programs: CONSENSUS and Gibbs sampling to the same set of sites. Using default parameters of these two programs we were not able to reveal two different patters. Only one pattern was revealed that presents a mixture of the original two: "**T(g/a)(c/a)GTCA**". It is noteworthy to pay attention that the kernel method by its nature is able to reveal correctly two matrices that are very similar to each other where most of other methods have much higher failure rate.

AhR. We have analyzed a relatively small set of 24 binding sites for transcription factor AhR that plays a very important role in the antitoxic cellular response. We identified two pattern subtypes: "TTGCGTGA"(matrix V$AHR_N1, see Fig.2) and "CTCGCGTG" (V$ANR_N2) that differ mainly at their 5' end and may correspond to two different groups of binding complexes.

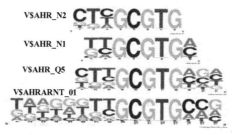

Fig. 2. Sequence logos of the weight matrices for AhR binding sites. Last two matrices are taken from TRANSFAC, first two are generated in this work.

3.3   Analysis of Alteration of Gene Expression in Human and Rat Hepatocytes by Toxins

We studied genes whose expression is regulated by a ligand-activated transcription factor AhR (aryl hydrocarbon receptor) that mediates responses to a variety of toxins. Expression of a number of genes was measured by RT-PCR in human and rat hepatocytes after treatment with Aroclor 1254 (artificial ligand of Ah – receptors). 111 (72 human and 39 rat) of differentially expressed genes were identified. To retrieve the promoter sequences of the genes we use Ensembl and DBTSS databases (Suzuki et al., 2002). The beginning of the annotated first exon was considered as a tentative TSS (transcription start site). For the analysis we selected the regions around TSS: -1000/+100.

To analyze the structure of these promoters we applied a novel method *CMFinder* (Kel et al, 2004 submitted). This method applies a genetic algorithm to identify so called composite modules (CMs) – specific combinations of TF patterns (in the form of weight matrices) that correlate with the level of up or down regulation of the genes. The program takes as input a full set of known TF matrices and computes an optimal combination of them that fits best to the observed expression changes of the genes.

We have included the new pattern subtypes for AhR binding sites that were found on the previous step of analysis. The result of the *CMFinder* program is shown in Figure 3. One can see that both new AhR patterns as well as some other previously constructed patterns were selected by the program. The composite module contained matrices for AhR, PPAR, HNF-6, STAT, ROR and ETS.

It is interesting that five different AhR matrices were included by the algorithm in the CM. This suggests that sites for AhR in the promoters play an important role in the up or down regulation of these genes. V$AHR_01 got the maximal impact value. It seems to be specific for several cytochrome P450 genes and influences their expression in response to AhR. V$AHRARNT_01 was constructed on the basis of data from SELEX experiments whereas V$AHR_Q5 was constructed on the basis of genomic sites. These two matrices as well as two matrices constructed by the Kernel method (V$AHR_N2 and V$AHR_N1) have very different impact values: Two of them (V$AHRARNT_01, V$AHR_N2) have a positive, the other two (VAHR_Q5, VAHR_N1) a negative impact value. Comparison of the structure of these two matrices shows that they are very similar in the core, but differ in some nucleotides at the flanks. For example, in position 10 of matrix V$AHR_Q5, the most prominent

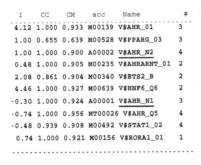
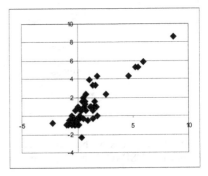

Fig. 3. Result of applying *CMFinder* program to the AhR gene expression data. A) The optimal composite module found by the program; two underlined matrices correspond to the two pattern subtypes of AhR sites; column I corresponds to the impact of the matrix into the correlation of the composite module score with the level of up or down regulation (positive impacting "up", negative – "down" regulation); CC, CM – core and matrix cut-offs; Acc and Name – TRANSFAC accession number and the name of the matrix. B) Plot of the correlation between composite module score (x-axis) and the log of gene expression change (y-axis).

nucleotide is G, whereas in the matrix V$AHRARNT_01 nucleotide "G" in the corresponding position is absolutely "forbidden". Matrices V$AHR_N1 and V$AHR_N2 are also different but on the other flank of the site. You can see in the Table 2 that the matrix V$AHR_N1 have got in the position 2 the consensus nucleotide "T" whereas in the matrix V$AHR_N2 the corresponding position is mainly occupied by nucleotide "C". This could influence binding of some other factors such as repressors in the vicinity of AhR sites.

In general, the Kernel method described in the paper is applicable to sets of unaligned regulatory sequences of any length. In a separate study we demonstrated its ability to reveal multiple patterns in the sets of tissue specific promoters.

Acknowledgments

This work was mainly funded by BIOBASE GmbH (Wolfenbüttel, Germany). Parts of this work were supported by Siberian Branch of Russian Academy of Sciences, by a grant of Volkswagen-Stiftung to A.K. (I/75941) and by a grant of the Lower Saxony Ministry of Culture and Science to J.B.

References

1. Bailey T.L., Elkan C. (1994) Proc Int Conf Intell Syst Mol Biol. 2, 28-36.
2. Buhler, J. and Tompa, M. (2002) J. Comput. Biol., 9, 225-242.
3. Ellrott K., Yang C., Sladek F.M., Jiang T. (2002) Bioinformatics. Suppl 2,S100-S109.
4. Hertz G.Z. and Stormo G.D., (1999) Bioinformatics, 15, 563-577.
5. Keich U., Pevzner P.A. (2002) Subtle motifs: defining the limits of motif finding algorithms. Bioinformatics,18,1382-1390.
6. Lawrence, C.E., Altschul, S.F., Bogouski, M.S., Liu, J.S., Neuwald, A.F., and Wooten, J.C., (1993) Science, 262, 208-214.

7. Pevzner, P.A. and Sze, S. (2000). In: Proceedings of the Eighth International Conference on Intelligent Systems for Molecular Biology, 269-278.
8. Thakurta, D.G. and Stormo, G.D. (2001) Bioinformatics, 17, 608-621.
9. Tikunov, Yu. and Kel, A.E. (2000) Kernel method for estimation of functional site local consensi. Classification of transcription initiation sites in eukaryotic genes. In: Proceedings of the German Conference on Bioinformatics (GCB00), October 5-7, 2000, Heidelberg, 83-88.
10. Tikunov, Y., Kel, A. (2004) Functional of averaged density: Application for estimation of probability density function. Submitted.
11. Workman, C.T. and Stormo, G.D. (2000) Pac. Symp. Biocomput., 5, 464-475.
12. Parzen E. (1962) On estimation of probability density function and mode, Annals of Mathematical Statistics, **33**, 1065-1076.
13. Rosenblatt M. (1956) Remarks on some nonparametric estimates of a density function, Annals of Mathematical Statistics, **27**, 832-837.
14. Orlov, A.I. (1991) Classification of nonnumeric objects on the basis of nonparametric density estimations. In: Problems of computer data analysis and modeling. Byelorussian State University, pp. 141-148.
15. Wingender, E., Chen, X., Fricke, E., Geffers, R., Hehl, R., Liebich, I., Krull, M., Matys, V., Michael, H., Ohnhäuser, R., Prüß, M., Schacherer, F., Thiele, S. and Urbach, S. (2001) The TRANSFAC system on gene expression regulation. Nucleic Acids Res. **29**, 281-283.

Learning Regulatory Network Models
that Represent Regulator States and Roles

Keith Noto and Mark Craven

Department of Biostatistics and Medical Informatics
Department of Computer Sciences
University of Wisconsin
Madison, Wisconsin 53706, USA
noto@cs.wisc.edu, craven@biostat.wisc.edu

Abstract. We present an approach to inferring probabilistic models of gene-regulatory networks that is intended to provide a more mechanistic representation of transcriptional regulation than previous methods. Our approach involves learning Bayesian network models using both gene-expression and genomic-sequence data. One key aspect of our approach is that our models represent *states* of regulators in addition to their expression levels. For example, the state of a transcription factor may be determined by whether a particular small molecule is bound to it or not. Our models represent these states using hidden nodes in the Bayesian networks. A second key aspect of our approach is that we use known and predicted transcription start sites to determine whether a given transcription factor is more likely to act as an activator or a repressor for a given gene. We refer to this distinction as the *role* of a regulator with respect to a gene. Determining the roles of a regulator provides a helpful bias in learning accurate representations of regulator states. We evaluate our approach using sequence and expression data for *E. coli* K-12. Our experiments show that our models are comparable to, or better than, several baselines in terms of predictive accuracy. Moreover, they have more explanatory power than either baseline.

1 Introduction

A significant, central challenge in computational biology is to develop methods that can elucidate biological networks from high-throughput data sources. In recent years, numerous research groups have developed methods that address the tasks of inferring regulatory [1] and metabolic networks [2] from data. Such models of biological networks can have both predictive and explanatory value. To achieve a high level of explanatory value, a model should represent the *mechanisms* of the network in as much detail as possible. In this paper, we describe an approach to inferring regulatory networks from gene-expression and genomic sequence data. Our approach incorporates several innovations that attempt to provide a more mechanistic representation than those used in previous work in this area. Our research has focused on prokaryotic genomes, and thus we empirically evaluate our method using sequence and expression data for *E. coli* K-12 [3]. Our experiments show that our models are able to provide expression predictions which are almost as accurate, and sometimes more accurate than several baselines with less explanatory value.

E. Eskin, C. Workman (Eds.): RECOMB 2004 Ws on Regulatory Genomics, LNBI 3318, pp. 52–64, 2005.

There are numerous factors that make the task of inferring networks from high-throughput data sources a difficult one. First, the available data characterizing states of cells, such as microarray data, are incomplete; they characterize the states of cells under a range of conditions that is usually quite limited. Second, there are typically high levels of noise in some of the available data sources, such as microarray and protein-protein interaction data. Third, measurements are not available for important aspects of the biological networks under study. For example, most efforts at network inference have employed only gene-expression measurements of protein-coding genes and genomic sequence data. However, in many cases gene regulation, even at the level of transcription regulation, is controlled in part by small molecules (*e.g.* IPTG inactivates the lac repressor), changes in protein states such as phosphorylation (*e.g.* arcA is activated through phosphorylation), or expression of small RNAs (*e.g.* 6S RNA associates with and regulates RNA polymerase).

Probabilistic models of gene regulation [4–13] are appealing because they can, in part, account for the uncertainty inherent in available data, and the non-deterministic nature of many interactions in a cell. The method that we present here builds on recent work in learning probabilistic graphical models to characterize transcriptional regulation.

Our approach, which involves learning Bayesian networks [14] using both gene-expression data from microarrays and genomic sequence data, incorporates several innovations. First, our models include hidden nodes that can represent the *states* of transcription factors. It is often the case that *expression levels* of transcription factors alone are not sufficient to predict the expression levels of genes they regulate. Transcription factors may not bind to a particular DNA site unless (or except when) they have bound a specific small molecule or undergone some post-translational modification. Given only microarray and genomic sequence data, we cannot directly measure theses states. However, we can think of these states as latent variables and represent them using hidden nodes in our Bayesian networks.

A second significant innovation in our approach is that we use known and predicted transcription start sites to determine whether a given transcription factor is more likely to act as an activator or a repressor for a given gene. We refer to this distinction as the *role* of a regulator with respect to a gene. To do this, we take advantage of a detailed probabilistic model of transcription units that we have developed in previous work [15]. Depending on the relative positions of a transcription factor binding site and a known or predicted promoter, we get an indication as to whether the transcription factor is acting as an activator or a repressor in a given case. We use this information to guide the initialization of parameters associated with the hidden nodes discussed above.

2 Approach

In this section, we first describe how we use Bayesian networks to represent various aspects of transcriptional regulation networks. A Bayesian network consists of two components: a qualitative one (the *structure*) in the form of a directed acyclic graph whose nodes correspond to the random variables, and a quantitative component consisting of a set of *conditional probability distributions* (CPDs). We then discuss how we learn both the structure and the parameters of our networks.

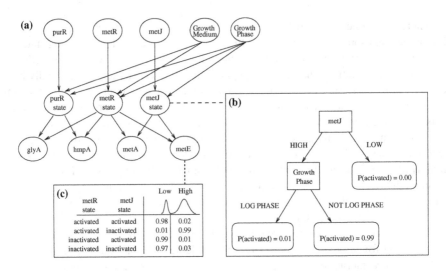

Fig. 1. (a) An example network with three regulators (purR, metR, and metJ), two cellular condition variables (Growth Medium and Growth Phase), and four regulated gene variables (glyA, hmpA, metA, and metE). (b) A possible CPD-Tree for the hidden node metJ-state. (c) A possible CPT for the regulatee node, metE, whose expression states are defined by a two-Gaussian mixture.

2.1 Network Architecture

Our models contain four distinct types of variables on three distinct levels. An example is shown in Fig. 1 (a). On the top level, there are nodes that represent the expression of regulators (genes whose products regulate other genes), and also nodes that represent the cellular conditions under which various gene-expression measurements were collected. On the bottom level, there are nodes representing the expression of genes known or predicted to be influenced by the regulators on the top level (we refer to these genes as *regulatees*). On the middle level, there are hidden nodes, one paired with each regulator node. These hidden nodes represent the "states" of the corresponding regulators. The parents of each hidden node are selected from a set of candidates that includes both the corresponding regulator expression node and the cellular condition nodes. The parents of each regulatee node are the hidden nodes corresponding to the regulators known or predicted to have a regulatory influence over that gene.

Each hidden node has two possible values, which can be interpreted as "activated" and "inactivated." As discussed in Section 1, regulators, such as transcription factors, are often activated or inactivated by *effectors*, such as small molecules. Although we do not have data that will allow us to directly detect the effectors for specific regulators, the network-learning algorithm can use cellular condition nodes as surrogates for these effectors. Consider for example, the transcription factor CAP which is activated by the small molecule cAMP. Our data do not contain cAMP measurements, but our method may learn that the absence of glucose in the growth medium is predictive of when CAP is activated. Thus the method has learned that glucose absence is a good surrogate for cAMP.

2.2   Representing Gene Expression States

We represent the expression levels of genes using a Gaussian mixture model [16]. We assume that most genes have multimodal expression-level distributions, with each mode corresponding to an "expression state" of the gene. Each Gaussian in the mixture represents the range of expression values for one state of the gene. Fig. 2 shows the mixture model inferred by our method for the metE gene.

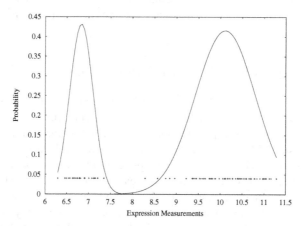

Fig. 2. Expression measurements for the gene metE, and a two-Gaussian mixture which describes its states. Expression measurements are plotted near the x-axis.

We use cross validation to choose the number of Gaussians in each mixture. Let $\mathbf{x}$ be the set of expression values for a given gene. For each fold, i, of cross validation, we divide $\mathbf{x}$ into two subsets, training data $\mathbf{x}'_i$ and held-aside data $\mathbf{x}''_i$. We use an expectation-maximization (EM) algorithm to optimize the parameters, Φ_i (the mean, variance and weight of each Gaussian in the mixture). However, we constrain the parameters so that the Gaussians are sufficiently far apart to ensure they each cover a separate range of expression values. Specifically, each Gaussian must have the highest probability density at each expression level within two standard deviations of its mean. The EM algorithm will attempt to optimize Φ_i as $argmax_{\Phi_i} P(\mathbf{x}'_i|\Phi_i)$, but only a local optimum is guaranteed. We then use the held-aside data to calculate the score for this fold as $score_i = P(\mathbf{x}''_i|\Phi_i)$. We repeat this process for one, two, or three Gaussians in a mixture. We choose the number of Gaussians associated with the highest score, $\sum_i score_i$, provided that a pairwise, two-tailed t-test determines the improvement to be statistically significant over the scores obtained from mixtures with fewer Gaussians. Once the number of Gaussians has been settled, we use the EM algorithm to optimize the final parameters Φ as $argmax_\Phi P(\mathbf{x}|\Phi)$, and consider each individual Gaussian to represent an expression state for that gene. If our method selects a mixture model with only one Gaussian for a given gene (*i.e.* there is only one expression state of this gene in the training set), then the gene is not included in the network model. In our experiments, this is the case for roughly half of the genes considered. About 90% of the remaining genes have two expression states, and 10% have three.

In order for our network to learn from data, we must express these data in terms of values of the variables in our network. For the regulator and regulatee nodes, these are expression states of the genes, but our evidence consists of expression measurements. Since the expression states are described by Gaussians over the expression values, it is straightforward to calculate a probability distribution over the expression states for each gene, given an expression measurement. These probability distributions are what we use as the evidence concerning the states of genes. Because of the constraints on the distance between Gaussians and the fact that they are positioned using the data, most such distributions will clearly favor a single Gaussian (gene expression state) over any other.

2.3   Representing Conditional Probability Distributions

As shown in Fig. 1 (b), we use trees to represent the conditional probability distributions for the hidden nodes in our networks. Each tree represents the distribution over values (*i.e.* states) of the corresponding hidden node, conditioned on the values of the node's parents. Recall that the candidate parents for each hidden node consist of the regulator expression level as well as the complete set of cellular condition nodes. We use trees to represent these CPDs for three key reasons. First, we assume that only a few of the candidate parents are relevant to modeling the regulator state, and thus we want the model to be able to select a small number of parents from a fairly large candidate pool. Second, trees provide descriptions of regulator states that are readily comprehensible and thus they can lend insight into the mechanisms which determine a regulator's behavior. Third, the trees can account for cases in which there is context-sensitive independence in determining a hidden node's probability distribution. In the sample tree shown in Fig. 1 (b), note how "Growth Phase" is only relevant if the regulator metJ has expression "HIGH." Note also how "Growth Medium" is not chosen as a parent at all.

Using our current data set, each regulatee in the network has a relatively small number of parents (between one and four), and we expect each parent to be relevant, so we use conventional *conditional probability tables* (CPTs) for the regulatee nodes, as shown in Figure 1 (c). The CPT for each regulatee node represents the distribution over the possible expression states of the particular gene, conditioned on the possible states of its parents.

2.4   Learning Network Parameters

Recall that each hidden variable is binary, and we refer to the possible values as "activated" and "inactivated". Since these states are unobserved, we cannot calculate the CPDs for the hidden and regulatee nodes directly. Instead, we set their parameters with an EM algorithm wherein we refine the CPDs iteratively until they converge to a local optimum which is consistent with the observed training data. Let $s_{R,i}^e$ represent the observed expression state of the ith regulator in experiment e, and let $s_{r,i}^e$ represent the observed expression state of the ith regulatee. Similarly, let $s_{c,i}^e$ denote the state of the ith cellular-condition variable in experiment e. We use Θ to represent all of the parameters of a Bayesian network, including the parameters (and structure) of each CPD tree and of each CPT. The EM algorithm adjusts the parameters trying to maximize the joint probability of the expression states across all experiments, given the cellular conditions:

$$\hat{\Theta} = \arg\max_{\Theta} \prod_e P(s^e_{R,1}, ..., s^e_{R,m}, \; s^e_{r,1}, ..., s^e_{r,n} \mid s^e_{c,1}, ...s^e_{c,k}, \; \Theta).$$

Here m is the number of regulators in the network, n is the number of regulatees, and k is the number of cellular-condition variables. We assume that we are always given the values of the cellular-condition variables, and thus our model represents the probability of the expression states conditioned on these values. The details of the E-step and M-step in this context are as follows.

E-Step: Let $S^e_{h,i}$ represent the (unobserved) state of hidden node i in experiment e (we use uppercase S to denote random variables in the Bayesian network and lowercase s to denote particular values the variables can take). In the E-Step, we compute the expected distribution over values of $S^e_{h,i}$ for each e and each i, given the observed expression states of the regulators and regulatees, and the observed cellular conditions:

$$P(S^e_{h,i} \mid s^e_{R,1}, ..., s^e_{R,m}, \; s^e_{r,1}, ..., s^e_{r,n}, \; s^e_{c,1}, ...s^e_{c,k}, \; \Theta).$$

This is computed as a Bayesian network query using *variable elimination* [17]. If the probability that a certain $S^e_{h,i}$ takes on the value "activated" is 0.7, then $S^e_{h,i}$ is treated as being 70% an "activated" data value and 30% an "inactivated" data value.

M-Step: Once these expected values are calculated, we use our now complete set of data to recalculate the network parameters. Let $\Theta_{h,i}$ refer to the CPD parameters for the ith hidden node and let $\Theta_{r,i}$ refer to the CPT parameters for the ith regulatee node. In the M-step, we attempt to maximize:

$$\prod_e P(s^e_{R,1}, ..., s^e_{R,m}, \; s^e_{r,1}, ..., s^e_{r,n}, \; s^e_{h,1}, ..., s^e_{h,k} \mid s^e_{c,1}, ...s^e_{c,k}, \; \Theta).$$

where $s^e_{h,1}, ..., s^e_{h,k}$ denotes the expectations for the hidden nodes calculated in the E-step. Each CPD-tree for hidden variable i, represented by $\Theta_{h,i}$, is re-grown by selecting a variable to split on (regulator expression state or cellular condition variable) which separates the set of expected values for the hidden variable, $\{s^e_{h,i}\}$, over all experiments e, into two subsets, $S^{\text{left}}_{h,i}$ and $S^{\text{right}}_{h,i}$, such that the classification error when considering the values in $S^{\text{left}}_{h,i}$ and $S^{\text{right}}_{h,i}$ to be either "activated" or "inactivated" is minimized. This process recurs on both subsets until no candidate split will further separate the data. A probability distribution over $S_{h,i}$ is then calculated for each leaf in the tree based on $\{s^e_{h,i}\}$ for the experiments e contained in that leaf's subset. Subtrees with a common ancestor that have nearly the same probability distributions over $S_{h,i}$ are pruned using an approach based on *minimum description length* (MDL), similar to one developed by Mehta *et al.* [18]. These pruned trees may not maximize the probability of the hidden states exactly, as do their unpruned counterparts. In practice, however, pruning speeds up convergence without sacrificing accuracy. Each regulatee CPT, $\Theta_{r,i}$, is also recalculated in the standard way during the M-step using $\{s^e_{r,i}\}$ and $\{s^e_{\pi(i)}\}$, where $\pi(i)$ is the set of parent nodes of regulatee node i.

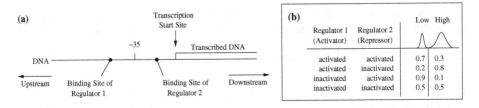

Fig. 3. (a) An example promoter configuration with one regulator binding site on each side of the -35 position. (b) A CPT for the gene that has been initialized based on the configuration of these two regulator binding sites.

2.5 Initializing Network Parameters

The EM algorithm described above will converge to a local optimum. In order to guide the network to converge to a good solution, we initialize the CPT for each regulatee based on prior knowledge about the roles of each its regulators. Specifically, for each regulatee we consider the relative location of the transcription start site, which is either known or predicted [15], and the binding sites for the regulators, which are also either known or predicted. We tentatively designate as *activators* those regulators that bind strictly upstream of the regulatee's promoter (which is estimated to extend 35 nucleotides upstream of its transcription start site), and we tentatively designate all other regulators as *repressors*. In the CPT for each regulatee, we assign a higher probability to the *highest* expression state when the putative activators are in the "activated" state, and we assign a higher probability to the *lowest* expression state when the putative repressors are in the "activated" state. We put more weight on this effect for repressors, which are believed to have a more stringent control on expression, and we put more weight on this effect when the regulatee's transcription start site is known than when it is only predicted. This initialization process is illustrated in Fig. 3.

3 Empirical Evaluation

In this section we present experiments designed to evaluate our Bayesian networks that (a) use hidden nodes to represent regulator states, (b) attempt to compensate for missing regulators, and (c) have the parameters associated with these nodes initialized to reflect their predicted roles (activator or repressor) with respect to individual genes.

3.1 Experimental Data and Methodology

We initialize the topology of our networks using 64 known and predicted *E. coli* regulators and their 296 known and predicted regulatees. The known instances are from TRANSFAC [19] and EcoCyc [20], and the predicted instances are based on binding sites predicted from cross-species comparison [21]. Our gene-expression data comes from a set of 90 Affymetrix microarray experiments [22]. Each array is annotated with experimental conditions, and the data are normalized using the *robust multiarray averaging* (RMA) technique [23].

We divide the microarray experiments into sets for which all of the annotated cellular conditions are identical (we call these replicate sets). From the original 90 experiments, there are 42 of these sets. The largest contains five experiments, and the others are copied so that each replicate set contains exactly five experiments. To assess model accuracy, we use leave-one-out cross-validation on each replicate set. That is, we hold one set of five identical experiments aside, train the model on the remaining experiments, and then evaluate the network on the held-aside data. For each testing example, we provide the network with expression levels of the regulators and the values of the cellular conditions, and then calculate a probability distribution over the possible expression states for each regulatee.

We evaluate the accuracy of our models using three measures. First, we calculate *classification error* as the extent to which the network predicts the incorrect expression states for each regulatee and experiment. Instead of calculating this error using "hard predictions" of the expression state of each regulatee, we take into account our uncertainty in each predicted expression state as well as the uncertainty in the discretization of each gene. In particular, we calculate classification error as follows:

$$\text{class error} = 100\% \times \left(1 - \frac{1}{E}\frac{1}{n}\sum_{e=1}^{E}\sum_{i=1}^{n}\sum_{d} P_{\Theta}(S_{r,i}^{e} = d)P_{\Phi}(S_{r,i}^{e} = d|x_{r,i}^{e})\right).$$

Here $P_{\Theta}(S_{r,i}^{e} = d)$ is shorthand for $P(S_{r,i}^{e} = d|s_{R,1}^{e}, ..., s_{R,m}^{e}, s_{c,1}^{e}, ..., s_{c,k}^{e}, \Theta)$ which is the probability, as predicted by the Bayesian network, that the ith regulatee is in state d for experiment e, given the expression values of the regulators and the cellular conditions. $P_{\Phi}(S_{r,i}^{e} = d|x_{r,i}^{e})$ represents the probability that the regulatee is in state d given expression measurement $x_{r,i}^{e}$, according to the Gaussian mixture model for this gene. Second, we compute the *average squared error*, where the error is the difference between the actual expression value and the means of the Gaussians representing each expression state, weighted by the predicted distribution over these states:

$$\text{ASE} = \frac{1}{E}\frac{1}{n}\sum_{e=1}^{E}\sum_{i=1}^{n}\sum_{d} P_{\Theta}(S_{r,i}^{e} = d)(\mu_{r,i,d} - x_{r,i}^{e})^{2}.$$

Here, $\mu_{r,i,d}$ is the mean of the Gaussian for state d in the mixture model for the ith regulatee. Third, we calculate the joint *log probability* of all test-set expression values, again taking into account our uncertainty in each predicted expression state and in discretization of each gene:

$$\text{log probability} = \sum_{e=1}^{E}\sum_{i=1}^{n}\log\left(\sum_{d} P_{\Theta}(S_{r,i}^{e} = d)P_{\Phi}(S_{r,i}^{e} = d|x_{r,i}^{e})\right).$$

We apply pairwise, two-tailed t-tests to test the statistical significance of differences between methods.

3.2 Experiment 1: The Value of Representing Regulator States

In order to test the value of including hidden nodes that represent regulator states, we compare against two baselines, examples of which are shown in Fig. 4. The first baseline employs Bayesian networks that have nodes representing the expression levels of

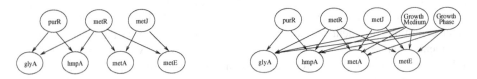

Fig. 4. Examples of the two baseline networks used in Experiment 1. These are the counterpart baselines for the network shown in Fig. 1.

regulators and regulatees, but which do not have hidden nodes representing regulator states. The second baseline augments the first by also incorporating cellular condition nodes, but it too does not have hidden nodes.

Table 1 shows the totals over all test folds for all three measures. For each of these measures, our models are more accurate than the baseline models which do not have hidden or cellular condition nodes. The differences are statistically significant at a confidence of over 99% for classification error and average squared error. The difference in overall probability is not statistically significant at a confidence of more than 95%. The baseline networks that include cellular condition nodes provide slightly more accurate predictions than our models with hidden nodes. However, we argue that these baseline models have a significant limitation in they do not provide a very mechanistic description of regulatee expression. That is, they do not directly represent the states of regulators and how these states govern the expression of regulatees. Thus, they have less explanatory power than our models.

Table 1. Predictive accuracy for the models with hidden nodes and the two baselines.

Model Variant	Classification Error	Average Squared Error	Log Probability
Full Model	16.59%	0.59	-12,066
Without Hidden Nodes	12.42	0.51	-12,193
Without Hidden or Cellular Condition Nodes	22.16	0.75	-13,363

Note also that our models show an improvement in overall log probability when compared to each of these baselines. Since the overall probability is a product of regulatee expression probabilities, an incorrect prediction with a probability very close to zero can have an unbounded effect on the final measurement. We hypothesize that our models make fewer of these extreme probability predictions because the regulatees are constrained by binary-valued parents.

3.3 Experiment 2: Discovering Missing Regulators

It is certainly the case that some relevant regulators are not represented in our networks. In this section, we consider a simple approach that dynamically adds hidden nodes to the networks. This approach tries to identify sets of regulatees for which a network makes incorrect predictions on many of the same training examples. After first training a network using the EM approach described earlier, we recursively cluster regulatees that share at least 50% of training examples incorrectly predicted by either regulatee (or cluster). A new hidden node is created for each cluster and this node becomes a parent of the regulatees in the cluster. The network is then re-trained using the EM approach. This procedure may be iterated a number of times.

Table 2 shows the resulting accuracy with up to three iterations of this procedure. Each iteration decreases the classification error of the models, and each decrease is statistically significant at a confidence above 97%. For the average squared error measure, only the decrease in error from the original model to any of the other three is statistically significant (at a confidence of 95% or greater). The application of this procedure improves the overall probability, but it does not continue to increase over multiple iterations. The differences in overall probability are not statistically significant.

Table 2. Predictive accuracy for the models with added hidden nodes.

Iterations	Classification Error	Average Squared Error	Log Probability
0 (Original model)	16.59%	0.59	-12,066
1	14.23	0.53	-11,586
2	13.65	0.51	-11,987
3	13.34	0.51	-12,004

Notice that this technique improves all three of our measurements, and that the classification error approaches that of the baseline without hidden nodes shown in Table 1, yet still provides models that explain relevant regulatory mechanisms.

3.4 Experiment 3: The Value of Initializing Regulator Roles

Recall that, before we train our network, we initialize the CPTs of the regulatee nodes based on the relative positions of known and predicted regulator binding sites and known or predicted promoters. We hypothesize that this initialization process will guide the EM algorithm toward a better solution. To evaluate the effectiveness of this technique, we compare the accuracy of our approach to a variant in which we initialize the parameters randomly. We also apply the technique of adding hidden nodes as described in the previous experiment because this increases the parameter space, and, one would expect, the importance of a good initialization.

The results of this experiment are shown in Table 3. Our initialization technique improved both the classification error and the average squared error, and the improvement

Table 3. Predictive accuracy for models with promoter-based parameter initialization and random initialization.

Initialization	Classification Error	Average Squared Error	Log Probability
Using Promoter Data	13.34%	0.51	-12,004
Random	14.19	0.54	-11,893

is statistically significant at a confidence above 96% for both measures. The technique did not improve the overall probability, however the decrease is not statistically significant. Repeating the experiment using random initialization many times on the same fold of cross validation, we estimate the standard deviation of the classification error at about 0.63% and of the average squared error at about 0.028. Notice that our initialization technique is an improvement over random initialization of at least this much. The standard deviation for log probability is estimated at about 25.53.

4   Conclusion

In addressing the problem of inferring models of transcriptional regulation, we have developed an approach that is able to learn to represent the states of regulators (*i.e.* whether a transcription factor is activated or not) as well as their roles (*i.e.* whether a transcription factor acts as an activator or repressor for a given gene). We have empirically evaluated our approach using gene-expression and genomic-sequence data for *E. coli* K-12. Our experiments show that both of these aspects of our approach result in models with a high level of predictive accuracy.

There are a number of extensions to our approach that we plan to investigate in future research. First, in keeping with our goal of learning more mechanistic representations, we plan to extend our models to account for additional types of regulatory mechanisms, such as riboswitches [24]. Second, we plan to adjust our approach so that we can relax some of the simplifying assumptions we have made in our initial work, such as the assumptions that genes have only one transcription start site, regulators have only one binding site in a given promoter region, and genes have distinct modes in their expression-level distributions. Third, we plan to extend our method so that the process of adding candidate regulators to a network involves looking for evidence of these regulators (*e.g.* transcription factor binding sites) in the genomic sequence. All of these proposed extensions are aimed at advancing the theme of learning models that exploit multiple sources of data, and attempt to provide mechanistic descriptions of regulatory relationships.

Acknowledgments

This research was supported in part by NLM training grant 5T15LM005359, NSF grant IIS-0093016, and NIH grant R01-LM07050-01. The authors would like to thank Joseph Bockhorst and Jesse Davis for helpful comments on an earlier draft of this paper.

References

1. de Jong, H.: Modeling and simulation of genetic regulatory systems: A literature review. Journal of Computational Biology **9** (2002) 67–103
2. King, R., Whelan, K., Jones, F., Reiser, P., Bryant, C., Muggleton, S., Kell, D., Oliver, S.: Functional genomic hypothesis generation and experimentation by a robot scientist. Nature **427** (2004) 247–252

3. Blattner, F.R., Plunkett, G., Bloch, C.A., Perna, N.T., Burland, V., Riley, M., Collado-Vides, J., Glasner, J.D., Rode, C.K., Mayhew, G.F., Gregor, J., Davis, N.W., Kirkpatrick, H.A., Goeden, M.A., Rose, D.J., Mau, B., Shao, Y.: The complete genome sequence of Escherichia coli K-12. Science **277** (1997) 1453–1474

4. Friedman, N., Linial, M., Pe'er, D.: Using Bayesian networks to analyze expression data. Journal of Computational Biology **7** (2000) 601–620

5. Hartemink, A., Gifford, D., Jaakkola, T., Young, R.: Using graphical models and genomic expression data to statistically validate models of genetic regulatory networks. In: Proceedings of the Fifth Pacific Symposium on Biocomputing, Kohala Coast, HI, World Scientific Press (2001) 422–433

6. Hartemink, A., Gifford, D., Jaakkola, T., Young, R.: Combining location and expression data for principled discovery of genetic regulatory networks. In: Proceedings of the Fifth Pacific Symposium on Biocomputing, Lihue, HI, World Scientific Press (2002) 437–449

7. Pe'er, D., Regev, A., Elidan, G., Friedman, N.: Inferring subnetworks from peturbed expression profiles. Bioinformatics **17** (2001) 215S–224S

8. Segal, E., Yelensky, R., Koller, D.: Genome-wide discovery of transcriptional modules from DNA sequence and gene expression. Bioinformatics **19** (2003) i273–i282

9. Segal, E., Shapira, M., Regev, A., Pe'er, D., Botstein, D., Koller, D., Friedman, N.: Module networks: Identifying regulatory modules and their condition-specific regulators from gene expression data. Nature Genetics **34** (2003) 166–176

10. Tamada, Y., Kim, S., Bannai, H., Imoto, S., Tashiro, K., Kuhara, S., Miyano, S.: Estimating gene networks from gene expression data by combining Bayesian network model with promoter element detection. Bioinformatics **19** (2003) ii227–ii236

11. Yoo, C., Cooper, G.: Discovery of gene-regulation pathways using local causal search. In: Proceedings of the Annual Fall Symposium of the American Medical Informatics Association. (2002) 914–918

12. Yoo, C., Thorsson, V., Cooper, G.: Discovery of causal relationships in a gene-regulation pathway from a mixture of experimental and observational DNA microarray data. In: Proceedings of the Fifth Pacific Symposium on Biocomputing, Lihue, HI, World Scientific Press (2002) 498–509

13. Ong, I., Glasner, J., Page, D.: Modelling regulatory pathways in E. coli from time series expression profiles. Bioinformatics **18** (2002) S241–S248

14. Pearl, J.: Probabalistic Reasoning in Intelligent Systems: Networks of Plausible Inference. Morgan Kaufmann, San Mateo, CA (1988)

15. Bockhorst, J., Qiu, Y., Glasner, J., Liu, M., Blattner, F., Craven, M.: Predicting bacterial transcription units using sequence and expression data. Bioinformatics **19** (2003) i34–i43

16. Xing, E., Jordan, M., Karp, R.: Feature selection for high-dimensional genomic microarray data. In: Proceedings of the Eighteenth International Conference on Machine Learning. (2001)

17. D'Ambrosio, B.: Inference in Bayesian networks. AI Magazine **20** (1999) 21–36

18. Mehta, M., Rissanen, J., Agrawal, R.: MDL-based decision tree pruning. In: Proceedings of the First International Conference on Knowledge Discovery and Data Mining, AAAI Press (1995) 216–221

19. Wingender, E., Chen, X., Fricke, E., Geffers, R., Hehl, R., Liebich, I., Krull, M., Matys, V., Michael, H., Ohnhäuser, R., Prüß, M., Schacherer, F., Thiele, S., Urbach, S.: The TRANS-FAC system on gene expression regulation. Nucleic Acids Research **29** (2001) 281–283

20. Karp, P., Riley, M., Saier, M., Paulsen, I., Collado-Vides, J., Paley, S., Pellegrini-Toole, A., Bonavides, C., Gama-Castro, S.: The EcoCyc database. Nucleic Acids Research **30** (2002) 56–58

21. McCue, L.A., Thompson, W., Carmack, C.S., Lawrence, C.: Factors influencing the identi-fication or transcription factor binding sites by cross-species comparison. Genome Research **12** (2002) 1523–1532

22. Glasner, J., Liss, P., III, G.P., Darling, A., Prasad, T., Rusch, M., Byrnes, A., Gilson, M., Biehl, B., Blattner, F., Perna, N.: ASAP, a systematic annotation package for community analysis of genomes. Nucleic Acids Research **31** (2003) 147–151

23. Irizarry, R., Hobbs, B., Collin, F., Beazer-Barclay, Y., Antonellis, K., Scherf, U., Speed, T.: Exploration, normalization, and summaries of high density oligonucleotide array probe level data. Biostatistics **4** (2002) 249–264

24. Nudler, E., Mironov, A.: The riboswitch control of bacterial metabolism. Trends in Bio-chemical Sciences **29** (2004) 11–17

Using Expression Data to Discover RNA and DNA Regulatory Sequence Motifs

Chaya Ben-Zaken Zilberstein[1], Eleazar Eskin[2], and Zohar Yakhini[3]

[1] Computer Science Dept., Technion
chaya@cs.technion.ac.il
[2] School of Computer Science and Engineering Hebrew University Jerusalem, Israel
eeskin@cs.huji.ac.il
[3] Agilent Laboratories and Computer Science Dept., Technion
zohar_yakhini@agilent.com

Abstract. The combination of gene expression data and genomic sequence data can be used to help discover putative transcription factor binding sites (TFBSs). There are two major approaches to incorporating expression data into the discovery of TFBS. The first approach clusters genes according to their expression patterns. Then, over-represented sequences are sought, in the promoter regions of co-expressed genes [31, 15, 14]. A second approach uses a single expression experiment and attempts to determine which transcription factors are involved in the experiment [24, 16, 29, 12].

In this paper, we present RIM-Finder, a further development of the second approach. Our method also enables the discovery of mRNA stability motifs and motif phrases. Phrases are either single motifs (a TFBS candidate or a RNA stability motif candidate) or pairs consisting of both types of motifs and a certain logical relation between them. Our approach discovers *all* (potentially degenerate) phrases that are statistically significant with respect to their distribution in a ranked list of sequences under either a non-parametric model or a Student t based model. In order to allow the identification of phrases consisting of both DNA and RNA motifs we rank sequence pairs consisting of promoters and mRNA untranslated regions (UTRs). We apply RIM-FINDER to discover putative phrases using cell stress response expression, mRNA decay rate measurements and mutant expression in yeast.

1 Introduction

Cellular levels of mRNA molecules are largely influenced by transcription rates as well as by degradation rates. The role of transcription factor binding sites (TFBSs) and therefore of promoter region sequence motifs is well recognized and well studied. Evidence is also emerging that links mRNA degradation processes to sequence motifs. These motifs specifically bind regulatory proteins or small regulatory RNA molecules and thus affect the stability of the mRNA molecule of the sequence in which they reside. For example, a recent study suggests that gene expression may be regulated, at least in part, at post-transcriptional level, by factors inducing extremely rapid degradation of mRNAs [2]. These factors include reactions between adenyl-uridyl-rich elements (AREs) of the target mRNA and specific proteins that bind to these elements.

E. Eskin, C. Workman (Eds.): RECOMB 2004 Ws on Regulatory Genomics, LNBI 3318, pp. 65–78, 2005.
© Springer-Verlag Berlin Heidelberg 2005

In this paper we suggest methods that facilitate the discovery of motifs that influence RNA stability by investigating whole genome expression profiling data and their relationship to promoters and RNA UTR sequences. We identify candidate motifs that contribute to RNA stabilization or de-stabilization.

Our methodology framework is based on searching phrases of motifs that are highly correlated with expression levels. These phrases are either single motifs (a TFBS candidate or a RNA stability motif candidate) or pairs consisting of both types of motifs and a certain logical relation between them. The logical relationship between DNA and RNA motifs in a phrase allows us to hypothesize functions. For example, using the union relation between an enhancer TFBS and a RNA stabilizer corresponds to each of these motifs being sufficient to induce high mRNA levels by itself. Another example is the intersection relation between a repressor TFBS and a RNA de-stabilizer which corresponds to both motifs being necessary in order to entail low mRNA levels. Motifs bases on logical phrases are also studied in [30] In order to identify significant phrases we consider pairs of sequences each consisting of the promoter and the 3'UTR of a single gene. These pairs are ranked according to gene expression. Ranking can also be induced by separation scores (for many experiments) [1, 20], ChIP on chip measurements [7], or any other relevant experimental results. The intuition behind our framework is that in either the highly over-expressed or highly under-expressed genes the sequence pairs will be enriched by hits of active phrases. We are therefore seeking motifs or phrases that occur in a rank-imbalanced manner in the ranked list of genes. A RIM (rank imbalanced motif) is a sequence motif which occurs with statistically significant frequency at either end of a ranked list, compared to its over-all frequency. The same intuition is also the basis of several previous studies that describe the use of a single expression experiment to facilitate the discovery of TFBSs. For example, Jensen and Knudsen, 2000[24] employed a non-parametric approach to discover motifs in the upstream regions of genes which are significant with respect to the ranking of the genes based on expression values for a single experiment. Bussemaker *et. al.* , 2001[16] presented the REDUCER algorithm which introduces the concept of motif regression where a linear model is assumed to underlie the expression data and the variables in the linear model are indicator variables corresponding to the presence of specific motifs. REDUCER also takes motif multiplicity into account when assessing a motif correlation with a set of expression values. Both of these approaches consider motifs as exact matching words. MOTIF-REGRESSOR, introduced in Conlon *et. al.* , 2003 [12], also assumes a similar linear model to REDUCER but improves by being able to find longer motifs and motifs with degenerate positions. MOTIF-REGRESSOR first identifies candidate motifs that are over represented in either the most over or under expressed genes and then uses the candidates to find the best motifs that correlate with expression. This method is shown, in the simulation experiments of [12], to find motifs in practice, but does not give any guarantees of finding the most significant motifs. Another related work is the approach of Keles *et. al.* , 2002[29] which also assumes a linear model of motif occurrence to expression levels and uses a feature selection method to discover the best motifs.

We score correlation between expression and phrase occurrence using both a non-parametric approach similar to the one used by [24] and a Student t approach which we developed for this purpose. To cope with the computational task involved in discovering significant phrases we have developed RIM-Finder, the core of which is an algorithmic framework which allows us to efficiently perform a search over large sets of phrases composed of pairs of motifs and the logical symbols NOT AND and OR. We employ efficient data structures to allow this search. In Appendix A we show that, under a certain formulation, the problem of finding RIMs is *NP-Complete*.

A confusion can occur between RNA and DNA motifs, when using our approach, especially if one uses 5'UTRs that are very close to the promoters and can contain active TFBSs. Thus, in order to make sure that our approach is able to identify motifs influencing RNA stability we used two different strategies. First, we applied RIM-Finder to yeast 3'UTRs and 5'UTRs and their measured decay rates. Since the mRNA decay rate is totally independent of the molecule transcription rate, the identified motifs can only be related to RNA stability. Second, we applied RIM-Finder to yeast expression data including stress and cell cycle, but we used only the 3'UTRs with promoters. In the latter case the identified RNA motifs can only influence RNA stability as in yeast most of the transcription regulation is known to occur in the promoters that are far enough from the 3'UTRs. Some significant RNA and DNA motifs as well as combined phrases were found in all of the above calculations. Lastly, to further validate our approach, we applied RIM-Finder to expression profiles of yeast mutants and of yeast cells with constituent over-expression; using promoter region sequences only. We show that the related TFBSs are successfully discovered.

2 Statistical Scores

2.1 Non-parametric Statistics

In order to evaluate the significance of a candidate phrase, we first rank the given genes by their expression levels and then label the genes as follows: genes whose promoters and UTRs contain the phrase are labeled by '+1' while the others are labeled by '-1'. One example is a phrase consisting of the union of a DNA motif and a RNA motif. In this case it is sufficient that the gene's promoter contains the DNA motif or that the corresponding UTR contains the RNA motif for the gene to be labeled by '+1'. If the given phrase is active under the studied condition, i.e. if it influences mRNA levels, then the '+1's are expected to have denser representation in one side of the labeled vector. On the other hand, if it is not active then the '+1's and '-1's will be interspersed. A phrase or a motif is said to be *rank imbalanced* if their occurrence vector is imbalanced in a statistically significant manner. We can also hypothesize the effect of the phrase and of each of it's motifs. For example, if the '+1's are over-represented at the top of this vector and the phrase is D OR NOT R where D is a DNA motif and R a RNA motif; we say that it induces high mRNA levels and that D is an enhancer and R is a de-stabilizer. However, if the '+1's are over-represented at the bottom we say that D is a repressor and R a stabilizer.

The *minimum hyper geometric* score (mHG) is a natural way to evaluate the significance of the given motif, based on partitions of the corresponding occurrences vector G. It corresponds to the partition that best divides $G \in \{+1, -1\}$ into a prefix and suffix, such that both are maximally homogeneous in terms of the symbols they contain. Formally, the mHG score of a vector G is defined as:

$$mHG(G) = \min_{x;y=G} HG(x, G) \qquad (1)$$

where $x; y$ denotes a partition of G into a prefix x and a suffix y and

$$HG(x, G) = \sum_{i=M_x}^{n} \frac{\binom{M}{i}\binom{|G|-M}{n-i}}{\binom{|G|}{n}}. \qquad (2)$$

Here $|G|$ represents the total number of genes, M is the total number of occurrences of the motif or the number of '+1's, M_x is the number of '+1's in the prefix x and n is the length of x. The p-values of the mHG score differ from the hyper geometric (HG) scores at which the mHG value is attained, since we must adjust for the multiple tests represented by different vector partitions. An upper bound on the p-value of any mHG level, can be obtained by multiplying it by the length of the vector, $|G|$:

$$\text{p-Val}\,(mHG(G) = s) \le |G|s. \qquad (3)$$

2.2 A Student t Approach

Unlike the non-parametric approach, which only takes into account the positions of the genes in the ranked list, this approach accounts for the actual expression levels of the genes. Moreover, when checking the significance of a single motif, it can be adjusted to take into account occurrence multiplicities. In this approach we check the significance of a phrase by performing a t-test to compare two distributions: the first, x_{phrase} is the distribution of the set of expression levels of genes containing the phrase. The second, x_{others} is the distribution of expression levels of all other genes. We adjust to multiple occurrences in the case of a single motif by adding gene expression levels to x_{phrase} multiplied by the number of times the motif occurs in relevant sequence, be it promoter or UTR. To evaluate the statistical significance of a motif we use a t-test as follows:

THEOREM Let x_1 and x_2 be two observations and let n_1 and n_2 be their sizes respectively. Let $\bar{x}$ denote the mean of sample x, and SD_x it's standard deviation. Let $SD = \sqrt{\frac{2SD_{x_1}(n_1-1)+2SD_{x_2}(n_2-1)}{n_1+n_2-2}}$. If x_1 and x_2 were sampled from the same normal population and assuming that $n_1 + n_2 \ge 30$ then $t = \frac{\bar{x_1}-\bar{x_2}}{SD\sqrt{1/n_1+1/n_2}}$ has an approximate standard normal distribution.

In order to check whether a phrase is active in a given condition we want to reject the hypothesis that x_{phrase} and x_{others} were sampled from the same population and, therefore, accept the hypothesis that the presence of the motif affects mRNA levels. Therefore, we calculate $\Phi(t)$, the corresponding level of significance associated with t. If it is significantly small and $\overline{x_{phrase}} \ge \overline{x_{others}}$ we conclude that the phrase induces high mRNA levels. On the other hand, if $\overline{x_{phrase}} \le \overline{x_{others}}$ we conclude that the phrase induces low mRNA levels.

3 Algorithmic Approach

The core of the approach is an efficient algorithm for performing the search through IUPAC patterns. For each IUPAC pattern, we determine exactly which sequences contain instances of a given IUPAC pattern and once we obtain the instances, we can compute either the mHG or Student t p-value. The algorithm for performing the search is a variant of the SPELLER algorithm described in Sagot, 1998[25].

The algorithm works on two *trie* data structures. A *trie* is a rooted tree with each edge labeled with a single symbol. The first is the *data trie* which contains a compressed representation of the data. The second is the *pattern trie* which represents the space of patterns in the search.

Consider the search for IUPAC patterns of length l. The data trie is created by considering all substrings in the data of length l extracted by a sliding window. The data trie is a rooted tree of depth l with each branch labeled with a nucleotide symbol $\{A, C, G, T\}$. The labels along the path from the root of the trie to the leaves correspond to the specific l-mer. Each leaf of the trie (depth l) corresponds to a specific l-mer and contains pointers to all occurrences of the l-mer in the data. Each internal node corresponds to all l-mers in the data which contain as a prefix the path from the root to the node. The data trie can be thought of as an index for the data. By following a path in the trie, we can efficiently recover the occurrences of any l-mer of interest. The trie is constructed as a preprocessing step to the motif search and can be constructed in linear time with respect to the total length of the sequence data.

The pattern trie corresponds to the space of IUPAC patterns. The pattern trie is a rooted trie of maximum depth l with each branch labeled with one of the IUPAC symbols in Table 1. Each leaf node (of depth l) corresponds to the IUPAC pattern of length l defined by the path from the root of the pattern trie to the leaf. A node in the pattern trie corresponds to the node in the data trie if the substring along the path from the root of the data trie to the data trie node matches the IUPAC pattern along the path from the root of the pattern trie to the pattern trie node. We will describe below how we construct the tree so that each leaf node contains a pointer to each leaf node in the data trie where the l-mer corresponding to the data trie leaf node matches the IUPAC pattern corresponding to the pattern trie leaf node. Using these pointers and the pointers from the data trie to the instances of the l-mers, we can recover all of the instances of substrings corresponding to the IUPAC pattern from the data. Traversing the entire pattern trie and checking the significance of each leaf node is equivalent to performing the full motif search.

The pattern trie is traversed and constructed in a depth first manner. The pattern trie is a *virtual* trie, i.e, only a single branch of the tree from the root to the current node is stored in memory. Each node in the pattern trie contains pointers to the corresponding nodes in the data trie at the same depth. Note that the root of the pattern trie contains a single pointer to the root of the data trie. The set of pointers for any node in the pattern trie can be efficiently derived from the set of pointers from its parent node by following all of the valid symbols with respect to the last symbol in the pattern from the nodes pointed to by the parents.

Consider the following example where $l = 2$. Initially, the pattern trie consists of only a root node pointing the root of the data trie. As we start the depth first traversal of the pattern trie, we construct the branch corresponding to A. The pointers for this node will consist of a single pointer to the node corresponding to A in the data trie since A is the only symbol that matches the IUPAC pattern. However, as the search progresses,

Table 1. IUPAC Alphabet for Nucleotide Sequences

Symbol	Meaning	Origin of Description
A	A	Adenine
C	C	Cytosine
G	G	Guanine
T	T	Thymine
M	A or C	aMino
R	A or G	puRine
W	A or T	Weak interaction (2 H bonds)
S	C or G	String interaction (3 H bonds)
Y	C or T	pYrimidine
K	G or T	Keto
V	A or C or G	not-T (not-U), V follows U in alphabet
H	A or C or T	not-G, H follows G
D	A or G or T	not-C, D follows C
B	C or G or T	not-A, B follows A
N	A or C or G or T	aNy

we will end up following the branch M and in that case the pointers will consist of the set pointing to nodes corresponding to A and C.

Since in the traversal we will only reach the IUPAC patterns that have at least one occurrence in the data, if we only consider the pattern alphabet without degenerate symbols our search embodies a linear time implementation of REDUCER.

Note that for the complete IUPAC pattern space, the number of leaf nodes that need to be traversed is upper bounded by 15^l which is impractical. However, in practice, the symbols that correspond to 3 nucleotides (V, H, D, B) are not very useful since they are captured in the combination of N and the symbols that represent 2 nucleotides (M, R, W, S, Y, K). In practice, we do not need to search over the entire 11^l space of patterns since *a priori* we know that very degenerate patterns are unlikely to have biological meaning. For our experiments we consider two IUPAC alphabets, a reduced degenerate alphabet consisting of the symbols $\{A, C, G, T, N\}$ and a full degenerate alphabet $\{A, C, G, T, M, R, W, S, Y, J, N\}$. Our implementation allows the user to define the maximal number of degenerate symbols in a pattern.

We extend the method to discover tandem motifs or motifs which have two conserved regions of length l separated by un-conserved spacing of length s. For example, consider motifs with 2 regions of 4 nucleotides separated by 8 nucleotides. Often, TF-BSs have this form since the TF's three dimensional structure drives it to bind in two locations which will be conserved, interspaced by a region that is not conserved since it is not involved in the actual binding. The only modification we need to make to the algorithm is that when constructing the data trie, we slide a window of length $2l + s$ and construct the trie out of substrings of length $2l$ representing the concatenation of the two conserved regions with the un-conserved spacing removed.

Logical motifs are discovered using the same data structure. The algorithm performs a depth first search over all patterns A and B. We first traverse the trie to search for pattern A. At each node in the search, we obtain a vector that records the sequences where pattern A occurs. We then traverse the trie again to discover B. At each node of this trie we obtain a vector that records where pattern B occurs. We then compute the occurrences of the logical motif by performing the appropriate operation on the two vectors. For example, if we are looking for the occurrences of the motif A AND B, we would perform an AND operation on the two vectors.

4   Results

4.1   RNA Stability Related Motifs

The abundance of each mRNA in the cell is determined not only by the rate at which it is produced, but also by it's decay rate. The decay rates of mRNA molecules can vary by 100-fold or more between different cell conditions [18, 10]. These rates are affected by a wide variety of stimuli and cellular signals, including: specific hormones [18, 19], iron [21, 4], cell cycle progression [26], cell differentiation [3, 17], and viral infection [9]. Proteins can effect mRNA decay rate by recognizing specific motifs in it. For example, many mRNA molecules with fast decay rates contain the sequence AUUUA in the 3'UTR, which de-stabilizes the mRNA by binding the cleavage and polyadenylation specificity protein factor (CPSF).

Genomewide determination of mRNA decay rates can be achieved by coupling a global transcriptional shut-off assay with DNA microarray analysis consisting of multiple temporal measurements of mRNA levels [33, 13]. Transcriptional shut-off is achieved by shifting the temperature of RNA polymerase II temperature sensitive mutants to 37°. A nonlinear least squares model can be fitted to estimate the decay rate k, of each mRNA. k is the value that minimizes $\sum_1^n \langle y(t_i) - exp(-kt_i) \rangle^2$, where $y(t)$ is the mRNA abundance at time t and the summation is taken over all time point observations. In Wang et al [33], for example, transcriptional shut-off was achieved by abruptly shifting the temperature from 24° to 37°, and microarray analysis was performed at 0, 5, 10, 15, 20, 30, 40, 50, and 60 min after the temperature shift.

We applied RIM-Finder to a list of 3'UTRS[1] ranked according to their decay rates [33]. The intuition is that an active stabilizer should occur more at the bottom of this list, therefore being rank imbalanced, and vice versa for an active de-stabilizer. The big advantage of using decay rates, is that the identified motifs must influence RNA stability and not transcriptional rate. Table 2 shows the results of our search.

4.2   Stress and Cell-Cycle Related Motifs

Next, we applied RIM-Finder to pairs consisting of promoters and 3'UTRs, ranked according to yeast stress expression data [6]. Very significant both RNA and DNA motifs as well as combined phrases were discovered. The strongest are represented in Table 3. Some motifs, e.g: GGGGA (STRE), AAATTTT, and GATGAG, appear with only

---

[1] We used the regions spanning 50bp downstream from the coding regions as 3'UTRS and those spanning 200bp upstream as promoters. To retrieve these sequences we combined data describing the chromosomal location of the coding sequences with chromosomal sequences. Both types of data were retrieved from the SGD database [23].

Table 2. Decay rate correlated RNA motifs. '+' represents a stabilizer; '-' a de-stabilizer. For Bonfferoni corrected p-values, entries of the table should be multiplied by 10^9

Motif's Sequence	log(mHG)	log(Student-t p-value)	Hypothesized Function
UGUNUANUA	-13	-7	-
GUSGUAW	-10	-7	+

small variations in most stress conditions. STRE is the well known multi-stress responsive element in yeast. Together with AAATTTT it has been reported, by previous studies, to be significantly correlated with expression or have such very close variant in cell cycle [24, 16], and in amino acid starvation [12]. GATGAG was also reported before to be significantly correlated with expression or have such very close variant in cell cycle [16] and in amino acid starvation [12]. The fact that these 3 motifs are significant in many different stress conditions derives us to hypothesize that they act as multi stress responsive elements, a hypothesis which is further strengthened by the fact that STRE has been empirically confirmed to induce expression under multi stress conditions [11, 22]. Moreover, GATGAG and AAATTTT have been shown to co-occur in a significantly high number of promoters in the S.cerevisiae, primarily in the promoters of genes involved in rRNA transcription and processing [27].

Since in yeast most of the transcription regulation occurs in the promoter areas, the RNA motifs discovered in the 3'UTRs are very likely to influence RNA stability. The RNA motif UGUNUANUA, which was found to be significantly correlated with decay rates, seems to have close variants which are significantly correlated with expression at several stress conditions. The fact that the same motif is discovered in several different experiments, strengthens our hypothesis regarding its effect in controlling RNA degradation.

Table 3 also contains significant phrases consisting of both DNA and RNA motifs. In most of these cases, the single motifs in the phrase would not be detected, due to insufficient statistical significance. For example, the phrase AGGG OR GTCNT is significant in heat-shock. However, GTCNT by itself gets a mHG score of only -3. Therefore, the increased flexibility of mixed phrases enables the discovery of more significant candidates.

We also applied RIM-FINDER to 18 lists of 3'UTRs each ranked according to expression at a different time point of the cell cycle [28]; several statistically significant motifs were discovered (data not shown). Moreover, when we calculated the mHG scores of motifs that were significant in one of the lists in all the other lists as well, we identified motifs that seem to be active at a limited number of time points, as reflected by their mHGs. These motifs seem to propel cell cycle by being a checkpoints between stages. For example, GTTGTTAC represented in Figure 1 has a mHG peak of 10^{-25} exactly at 49 minutes and therefore, seems to be important in controlling the cell transition from G2 to Miosis. This motif might bind a cyclin that acts by regulating degradation. Cyclins are proteins that are known to propel the cell cycle by transcriptional control and are active only at time points between stages. Our findings suggest that cyclins might propel the cell cycle also by controlling degradation.

Table 3. Single Motifs and Phrases influencing cellular mRNA levels under different environmental stress conditions. '+' represents a positive effect on mRNA levels and '-' indicates the opposite. For Bonfferoni corrected p-values, entries of the table should be multiplied by 10^{11}

Stress	Motif Type	Motif's Sequence	log(mHG)	Student-t log(p-value)	Hypothesized Function
Heat Shock	DNA	MGATGAG	-57	-50	-
5 min.	DNA	AAATTTT	-43	-48	-
	DNA	CTCATCK	-33	-31	-
	DNA	RGGGG	-27	-29	+
	RNA	UGUAUANUA	-15	-14	-
	RNA	GUAYAWU	-11	-14	-
	RNA	UUYUNSC	-10	-5	+
	Phrase	DNA motif AGGG or RNA motif GUCNU	-13	-8	+
	phrase	DNA motif CGNGG or RNA motif UUUUU	-12	-9	+
Sorbitol Shock	DNA	AAAATTT	-65	-63	-
10 min.	DNA	MSATGAG	-34	-38	-
	DNA	SKCATCG	-24	-28	-
	DNA	WRGGG	-20	-22	-
	RNA	UGUAUANUA	-11	-13	-
	phrase	CCCTT DNA motif or UUUNU RNA motif	-12	-10	-
Amino Acid	DNA	AAATTTY	-38	-23	-
Starvation 60 min.	DNA	YRTATAA	-27	-19	+
	DNA	CGATGMS	-21	-16	-
	DNA	TGAAWARA	-19	-5	-
	DNA	YNNKNC	-19	-14	-
	RNA	UCUAUNACA	-15	-12	+
	RNA	GUUGGANUA	-14	-15	-
	RNA	GNUGGUAUG	-13	-11	-
	RNA	CAUUMYG	-12	-9	+
	RNA	UGGKUGG	-12	-12	+
	phrase	GNGGA DNA motif or TNTTT RNA motif	-10	-9	+
Diamide Shock	DNA	AAAATTT	-73	-63	-
10 min.	DNA	MGATGAG	-72	-70	-
	DNA	CTCATCK	-43	-38	-
	DNA	GCGMTS	-29	-23	-
	DNA	RGGGR	-28	-28	+
	RNA	UNNNUAUAU	-12	-5	+
	RNA	UNYWUNU	-11	-4	+
	phrase	ANGGG DNA motif or UU CA RNA motif	-13	-10	+

4.3   Saccharomyces Cerevisiae Mutants and Related TFBSs

Finally, we applied RIM-FINDER to promoters ranked according to expression of different mutants or of cells constituently over-expressing a certain gene. We sought to identify the TFBSs of the related TF's as significant. The strongest motifs discovered, for each such experiment are described by Table 4.

GCN4 and STE12 are both positive regulators of transcription in yeast. We found close variants of their known binding sites in the list of the 8 strongest motifs discovered in their mutants: TGASTMA which represents a subset of TGANT, the binding site of GCN4, was found as most significant in GCN4 mutants; while, TGMAACR which includes TGAAACA, the known binding site of STE12, was found to be the 8th most significant motif in STE12 mutants. Moreover, as expected, both these motifs were over-represented at the bottom of the expression rank list.

MSN2 induces transcription in yeast, while ROX1 represses it. In MSN2 over-expressing cells we found CCCCT, the compliment sequence of STRE, the known binding site of MSN2, as the fifth strongest motif. Moreover, as expected, CCCCT

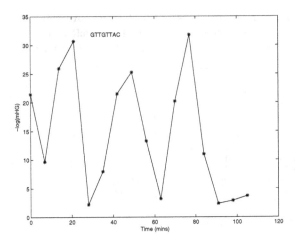

Fig. 1. GTTGTTAC a cell-cycle propeling motif. Stages of the S.cerevisiae cell-cycle: M/G1: 0-14 mins and 63-77 mins, G1: 14-21 mins and 77-91 mins, S: 21-42 mins and 91-105 mins, G2: 42-49 mins, and M: 49-63 mins

Table 4. Motifs that influence gene expression in yeast mutants. The motifs similar to the binding sites of the mutated proteins are marked by *. '+' represents a transcription enhancing influence, while, '-' a repression. For Bonfferoni corrected p-values, entries of the table should be multiplied by 10^{11}

Mutant	Motif Sequence	log(mHG).	log(Student-t p-value)	Hypothesized Function
GCN4 mutants	TGASTMA*	-29	-50	-
grown in complete	TATAWAW	-27	-10	-
medium [32]				
STE12 mutants	YNNKNC	-23	-31	+
grown in complete	GGAKTCC	-19	-21	-
medium compare	CCTYGAS	-16	-12	-
to WT cells	GSRAGCT	-16	-12	+
grown in the same	TAATAGG	-15	-18	-
medium [32]	WMAG	-15	-12	+
	GATMMTG	-13	-8	-
	TGMAACR*	-13	-13	-
	YTNNYT	-13	-10	+
ROX1 overexpressing	ACAATR	-30	-28	-
cells grown in	ATTGTY*	-21	-23	-
complete medium	ANTKKTT	-13	-12	-
compare to WT				
grown in the same				
medium [5]				
MSN2 overexpressing	MNGGRG	-25	-21	+
cells grown in	CCCST*	-19	-14	+
complete medium	MNCCCST*	-19	-18	+
compare to WT	SGGGNNS*	-19	-19	+
grown in the same	CCCCT*	-17	-19	+
medium[6]	ATATAAR	-17	-10	+

was over-represented in the top of the expression rank list. In ROX1 mutants we found a variant of ROX1 binding site: ATTGTY which is close to YYNATTGTTY as the second strongest motif. This motif was over-represented in the bottom of the expression ranked list. These findings are consistent with our expectations: since MSN2 is an enhancer, in the MSN2 over-expressing cells we expect the genes regulated by MSN2 to be over-expressed thereby making STRE over-represented in the highly expressed promoters; on the flip side, since ROX1 is a repressor, in the ROX1 over-expressing cells we expect the genes regulated by ROX1 to be under-expressed, thereby making ROX1 binding site over-represented in the low expression promoters.

5   Summary and Discussion

We have presented a new framework for discovering motifs that are significant with respect to a single gene expression measurement assay or to any other ranked list of genes. We evaluate motifs using a Student-t as well as a non-parametric model. We assign p-values for denser occurrence at either end of such lists. Our approach guarantees the discovery of the most significant degenerate motifs, an advantage over previous methods such as REDUCER or MOTIF-REGRESSOR.

We also applied our methods to RNA decay rates to produce motifs that may be active in regulating degradations. In particular, we found a RNA motif that seems to be a checkpoint between cell-cycle stages. Thus, we provide evidence that cell-cycle is partly propelled by controlling degradation rates.

References

1. Ben-Dor A, Friedman N, and Yakhini Z. Scoring genes for relevance. URL: http://www.labs.agilent.com/resources/techreports.html. Technical report, 2002.

2. Bevilacqua A, Ceriani MC, Capaccioli S, and Nicolin A. Post-transcriptional regulation of gene expression by degradation of messenger RNAs. *J Cell Physiol*, 195(3):356–372, 2003.

3. Krowczynska A, Yenofsky R, and Brawerman G. Regulation of messenger RNA stability in mouse erythroleukemia cells. *J Mol Biol*, 181(2):231–239, 1985.

4. Thomson AM, Rogers JT, and Leedman PJ. Iron-regulatory proteins, iron-responsive elements and ferritin mRNA translation. *Int J Biochem Cell Biol*, 31(10):1139–1152, 1999.

5. Gasch AP, Huang M, Metzner S, Botstein D, Elledge SJ, and Brown PO. Genomic expression responses to DNA-damaging agents and the regulatory role of the yeast ATR homolog Mec1p. *Mol Biol Cell*, 12(10):2987–3003, 2001.

6. Gasch AP, Spellman PT, Kao CM, Carmel-Harel O, Eisen MB, Storz G, Botstein D, and Brown PO. Genomic expression programs in the response of yeast cells to environmental changes. *Mol. Biol. Cell*, 11(12):4241–4257, 2000.

7. Ren B, Robert F, Wyrick JJ, Aparicio O, Jennings EG, Simon I, Zeitlinger J, Schreiber J, Hannett N, Kanin E, Volkert TL, Wilson CJ, Bell SP, and Young RA. Genome-wide location and function of DNA binding proteins. *Science*, 290:2306–2309, 2000.

8. Linhart C and Shamir R. The degenerate primer design problem. *Bioinformatics*, 18:172–181, 2002.

9. Sorenson CM, Hart PA, and Ross J. Analysis of herpes simplex virus-induced mRNA destabilizing activity using an in vitro mRNA decay system. *Nucleic Acids Res.*, 19:4459–4465, 1991.

10. Cabrera CV, Lee JJ, Ellison JW, Britten RJ, and Davidson EH. Regulation of cytoplasmic mRNA prevalence in sea urchin embryos. Rates of appearance and turnover for specific sequences. *J Mol Biol*, 174(1):85–111, 1984.

11. Moskvina E, Schuller C, Maurer CT, Mager WH, and Ruis H. A search in the genome of Saccharomyces Cerevisiae for genes regulated via stress response elements. *Yeast*, 11:1041–1050, 1998.

12. Conlon EM, Liu XS, Lieb JD, and Liu JS. Integrating regulatory motif discovery and genome-wide expression analysis. *Proc Natl Acad Sci U S A.*, 18;100(6):3339–3344, 2003.

13. Holstege FC, Jennings EG, Wyrick JJ, Lee TI, Hengartner CJ, Green MR, Golub TR, Lander ES, and Young RA. Dissecting the regulatory circuitry of a eukaryotic genome. *Cell*, 95:717–728, 1998.

14. Roth FP, Hughes JD, Estep PW, and Church GM. Finding DNA regulatory motifs within unaligned noncoding sequences clustered by whole-genome mRNA quantitation. *Nat Biotechnol*, 16(10):939–945, 1998.

15. Bussemaker HJ, Li H, and Siggia ED. Building a dictionary for genomes: identification of presumptive regulatory sites by statistical analysis. *Proc Natl Acad Sci U S A*, 97(18):10096–10100, 2000.

16. Bussemaker HJ, Li H, and Siggia ED. Regulatory element detection using correlation with expression. *Nat Genet*, 27(2):167–171, 2001.

17. Jack HM and Wabl M. Immunoglobulin mRNA stability varies during B lymphocyte differentiation. *EMBO J.*, 7(4):1041–1046, 1988.

18. Ross J. mRNA stability in mammalian cells. *Microbiol Rev*, 59(3):423–450, 1995.

19. Ross J. Control of messenger RNA stability in higher eukaryotes. *Trends. Genet.*, 12(5):171–175, 1996.

20. Storey JD and Tibshirani R. Statistical methods for identifying differentially expressed genes in dna microarrays. *Methods Mol Biol*, 224:149–157, 2003.

21. Casey JL, Hentze MW, Koeller DM, Caughman SW, Rouault TA, Klausner RD, and Harford JB. Iron-responsive elements: regulatory RNA sequences that control mRNA levels and translation. *Science*, 240:924–928, 1988.

22. Treger JM, Magee TR, and McEntee K. Functional analysis of the stress response element and its role in the multistress response of Saccharomyces cerevisiae. *Biochem Biophys Res Commun*, 243(1):13–19, 1998.

23. Dolinski K, Balakrishnan R, Christie KR, Costanzo MC, Dwight S S, Engel SR, Fisk DG, Hirschman JE, Hong EL, Issel-Tarver L, Sethuraman A, Theesfeld CL, Binkley G, Lane C, Schroeder M, Dong S, Weng S, Andrada R, Botstein D, and Cherry JM. Saccharomyces genome database. ftp://ftp.yeastgenome.org/yeast/.

24. Jensen LJ and Knudsen S. Automatic discovery of regulatory patterns in promoter regions based on whole cell expression data and functional annotation. *Bioinformatics*, 16(4):326–333, 2000.

25. Sagot M. Spelling approximate or repeated motifs using a suffix tree. *Lecture Notes in Computer Science*, pages 111–127, 1998.

26. Heintz N, Sive HL, and Roeder RG. Regulation of human histone gene expression: kinetics of accumulation and changes in the rate of synthesis and in the half-lives of individual histone mRNAs during the hela cell cycle. *Mol Cell Biol*, 3(4):539–550, 1983.

27. Sudarsanam P, Pilpel Y, and Church G. Genome-wide co-occurrence of promoter elements reveals a cis-regulatory cassette of r-RNA transcription motifs in S. cerevisiae. *Genome Research*, 12:1723–1731, 2002.

28. Spellman PT, Sherlock G, Zhang MQ, Iyer VR, Anders K, Eisen MB, Brown PO, Botstein D, and Futcher B. Comprehensive identification of cell cycle-regulated genes of the yeast Saccharomyces cerevisiae by microarray hybridization. *Mol Biol Cell*, 9(12):3273–3297, 1998.

29. Keles S, van der Laan ML, and Eisen MB. Identification of regulatory elements using a feature selection method. *Bioinformatics*, 18:1167–1175, 2002.

30. Kaminer T, Laor N, Lipson D, and Yakhini Z. Applying GAs in searching motif patterns in gene expression data. URL: http://bioinfo.cs.technion.ac.il/projects/kaminer-laor. Technical report, 2003.

31. Guha TD, Palomar L, Stormo GD, Tedesco P, Johnson TE, Walker DW, Lithgow G, Kim S, and Link CD. Identification of a novel cis-regulatory element involved in the heat shock response in Caenorhabditis elegans using microarray gene expression and computational methods. *Genome Res.*, 12(5):701–712, 2002.

32. Hughes TR, Marton MJ, Jones AR, Roberts CJ, Stoughton R, Armour CD, Bennett HA, Coffey E, Dai H, He YD, Kidd MJ, King AM, Meyer MR, Slade D, Lum PY, Stepaniants SB, Shoemaker DD, Gachotte D, Chakraburtty K, Simon J, Bard M, and Friend SH. Functional discovery via a compendium of expression profiles. *Cell*, 102:109–126, 200.

33. Wang Y, Liu CL, Storey JD, Tibshirani RJ, Herschlag D, and Brown PO. Precision and functional specificity in mRNA decay. *Proc Natl Acad Sci U S A*, 99(9):5860–5865, 2002.

Appendix

A   Computational Complexity

In order to identify new active TF binding sites, we seek motifs with significant mHG. In this section we prove that the problem involved in this task is *NP-complete*. We do so by showing that the decision version of this problem, the RANK IMBALANCED MOTIFS, is also *NP-complete*. We start by introducing some notations following Linhart and Shamir [8].

Let Σ denote a finite fixed alphabet. In the case of DNA sequences, $\Sigma = \{A, C, G, T\}$. A degenerate motif is a string P with several possible characters at each position, i.e., $P = p_1 p_2 \ldots p_k$, where $p_i \subseteq \Sigma$ (a motif over the IUPAC alphabet is any degenerate motif over $\{A, C, G, T\}$). A string $S = s_1 s_2 \ldots s_l, s_i \in \Sigma$ matches the degenerate motif P, if it contains a sub-string that can be extracted from P by selecting a character at each position, i.e. $\exists j, 0 \leq j \leq l - k$ such that $\forall i, 1 \leq i \leq k, s_{j+i} \in p_i$. For example, the motif $P^* = \{A\}\{C, G\}\{A, T, G\}$ matches the string TGAGAGTC starting from the third position. The *degeneracy* of P is $d(P) = \prod_{i=1}^{k} |p_i|$. For example, $d(P^*) = 6$.

Recall that the functionality of a motif is related to the mHG of its corresponding occurrence vector, given an expression ranked list of promoters. For a degenerate motif this vector is simply calculated by replacing strings that match the motif by '+1' and the others by '-1'.

Therefore, the computational problem of discovering new TF-binding sites, is related to the following decision problem.

PROBLEM1 (RANK IMBALANCED MOTIFS (RIMS)) Given a ranked set of n strings, R, over an alphabet Σ, integers l, d and a constant p; is there a degenerate motif P of length l, such that $d(P) \geq d$ and such that it's corresponding vector with respect to R, has a mHG $\leq p$

THEOREM RIMS is *NP-complete* for $|\Sigma| \geq 3$. We show this by showing a reduction from the Set-Cover problem, since we could not find any simple reduction from any of the 3 basic NP-Complete problems represented by Linhart and Shamir [8].

However, two special cases where the problem is solvable in polynomial time, should be pointed out. The first is where $d = 1$, and where all the possible solutions are sub-strings of the data. The second is where the motif length (l) is constant, and therefore, the number of possible solutions is also constant being $2^{|\Sigma|^l}$. In that case, one can solve the problem in time linear in the size of the data, by calculating the score of each of the motifs.

A.1   Reduction from Set-Cover

We use the vector representation of the set cover problem: consider a set of subsets $\{s_1, s_2 \ldots s_n\}$ of the universal set $U = \{1, 2, \ldots m\}$. Each subset is represented by a vector of length m, and the ith cell of V_j equals 1 *iff* $i \in s_j$ and 0 otherwise. We seek the smallest subset T, of vectors, such that each position between 1 to m is covered by at least one of the vectors in T. In the decision version we want to determine whether such cover of cardinality k exists.

1. Let $\Sigma = \{0, 1, 3\}$.
2. Let $M = v_1 \ldots v_n$ be a matrix, whose columns are the given vectors of the set-cover problem.
3. Let α be a sequence consisting of all the rows in M chained together, with the symbol 3 between them.
4. Let $\gamma = 30^{n+2}3 \circ \alpha \circ 30^n$ and $\beta = \gamma \circ 3$.
5. Let $R = \{\beta\} \cup \{\gamma\}$ and let β be ranked before γ in R.
6. Set l=n+2.
7. Set $d = 4 \cdot 3^{(n-k)}$
8. Set $p = 0.5$

A.2   Correctness of the Reduction

We now show, that there is a k cover *iff* there is a motif P of length l and $d(P) \geq d$ with score smaller than 1. The correctness is based on the fact that if $v = (+, +)$, its mHG is 1. While, if $v = (+, -)$ it's mHG is 0.5.

The first direction: Given a k cover, P is the motif of length $n + 2$ starting and ending with $\langle 3, 1 \rangle$ and having 0s in the positions corresponding to the cover and $\langle 0, 1, 3 \rangle$ elsewhere. This motif occurs in β but not in γ (it occurs in 30^n3 and does not occur in α) by construction.

The second direction: assuming that a motif of length $n + 2$, and degeneracy $\geq d$ got scores of 0.5. Then, it must occur in β in 30^n3 and it must start and end with $\langle 3 \rangle$ or $\langle 3, 1 \rangle$, otherwise it would occur in $30^{n+2}3$ and therefore, in γ. Since it has a degeneracy $\geq 4 \cdot 3^{(n-k)}$ it cannot have more than k zeros. Moreover, it is easy to show that since this motif does not occur in α its zeros must correspond to a k cover.

Parameter Landscape Analysis
for Common Motif Discovery Programs

Natalia Polouliakh[1,2], Michiko Konno[2], Paul Horton[3], and Kenta Nakai[1]

[1] Human Genome Center, University of Tokyo, 4-6-1 Shirokanedai, Tokyo, Japan
[2] Graduate School of Humanity and Science, Ochanomizu University,
2-1-1 Otsuka, Tokyo, Japan
[3] National Institute of Advanced Industrial Science and Technology, 2-43 Aomi, Tokyo, Japan

Abstract. The identification of regulatory elements as over-represented motifs in the promoters of potentially *co*-regulated genes is an important and challenging problem in computational biology. Although many motif detection programs have been developed so far, they still seem to be immature practically. In particular the choice of tunable parameters is often critical to success. Thus knowledge regarding which parameter settings are most appropriate for various types of target motifs is invaluable, but unfortunately has been scarce. In this paper, we report our parameter landscape analysis of two widely-used programs (the Gibbs Sampler (GS) and MEME). Our results show that GS is relatively sensitive to the changes of some parameter values while MEME is more stable. We present recommended parameter settings for GS optimized for four different motif lengths. Thus, running GS four times with these settings should significantly decrease the risk of overlooking subtle motifs.

1 Introduction

One of the central challenges in modern biology is to elucidate the gene regulatory networks of various organisms. To this end, a large amount of systematic gene expression data is accumulating. A typical way to explore such data is to find potentially *co*-regulated genes, i.e., genes that are regulated by a common transcription factor, by clustering genes showing similar expression patterns [1]. The next step is to find potential transcription factor-binding sites (*cis*-elements) from their upstream sequences (promoters) as over-represented motifs because the found motifs can be validated experimentally and because these motifs are also useful for finding other *co*-regulated genes [2]. However, finding such motifs is far from a trivial task. One difficulty is due to the fact that motif occurrences exhibit considerable variety, typically only partially matching the consensus pattern for the motif they belong to. Another difficulty is that there are many patterns in the genome, for example various repetitive elements, which are not transcription factor binding sites – and thus are a source of false positives for this task.

Although many motif finding programs have been constructed so far [3–12] the problem has not been satisfactory solved in a practical sense, especially for the analysis of higher eukaryotic promoters. To overcome this difficulty, what is needed first is a precise assessment of the capability of existing algorithms in various situations.

E. Eskin, C. Workman (Eds.): RECOMB 2004 Ws on Regulatory Genomics, LNBI 3318, pp. 79–87, 2005.

Such attempts have been extensively done by Pevzner's group [13]. In their first attempt, artificial motifs of a fixed length with a fixed number of mutations were embedded in a set of random DNA sequences with a fixed length for input sequences (the FM model; [13]). They tabulated the maximum number of tolerated mutations for each motif length. Later, real motifs were embedded for a more realistic assessment [14]. One of their conclusion for users was "use all available programs" because the relative superiority of each program varies with various situations[14]. Not only are several motif discovery programs available, but most of them have tunable parameters, which can significantly affect their performance. The Melina program alleviates this problem somewhat by conveniently allowing users to compare the results of four programs applied to the same data with various parameter settings [15]. Still, end-users need some kind of guidance to choose the right parameter values for their data but such guidance has not been readily available. In this paper we use Pevzner's scheme to conduct a parametric analysis of two easily-obtainable and widely used programs, the Gibbs sampler (GS) [5] and MEME [4, 6, 7]. In addition to providing a practical resource for end-users we believe our results may give valuable hints to algorithm designers as well.

2   Methods

2.1   Dataset Construction

To analyze the distribution of mutations and lengths of actual motifs, we ran-domly selected 200 groups of known eukaryotic motifs from the TRANSFAC database [16] and 104 groups of prokaryotic motifs from the DBTBS database [17]. For our experiments, we generated datasets analogous to those used in previous evaluations [13]. Motifs with lengths between 8 and 20 nucleotides were incorporated in a randomly generated background (1/4 probability of each nucleotide). The number of background sequences in each dataset was initially 30 and their sequence length was set to 600 nucleotides. Mismatches in motifs were inserted according to the original FM model [13]. We also used their result on the minimum tolerated number of mismatches (Table 1). We call their mismatch level "level one" and define "level two" as one more mismatch than level one for each motif length. Since it is quite common in practice that some of the input sequences do not contain motifs, we also constructed datasets by adding 30, 60 and 90 random sequences to the above 30 sequences.

Table 1. Motifs with corresponding number of mismatches used for the datasets construction.

motif	8	9	10	11	12	13	14	15	16	17	18	19	20
mismatches (level 1)	1	1	1	2	2	3	3	3	4	4	5	5	6
mismatches (level 2)	2	2	2	3	3	4	4	4	5	5	6	6	7

2.2   Evaluation of Elucidation Performance (EP)

We evaluated the program's sensitivity (*Sn.*) and specificity (*Sp.*), which are shown in the formulas below:

$$Sn. = TP/(TP + FN) \tag{1}$$
$$Sp. = TP/(TP + FP) \tag{2}$$

where TP (True Positive) corresponds to the number of perfectly found inserted motifs and of reported segments that have the same or lower number of mismatches; FN (False Negative) refers to the missed incorporated motifs. And FP (False Positive) corresponds to the reported segments which are not true positives. We obtained our estimates for sensitivity and specificity by averaging over 15 randomly generated datasets for each motif length and mismatch level. When optimizing parameters, we choose to use the product of sensitivity and specificity as the quantatity to maximize.

3   Results

3.1   Range of Mismatches and Lengths in Actual Motifs

When looking for biological motifs, we need to know something about their general features. For this purpose, we investigated 200 groups of eukaryotic motifs stored in the TRANSFAC [16] database and 104 groups from DBTBS (DataBase of Transcription regulation in Bacillus subtilis) [17]. It was determined that 99% of biological motifs have length between 5 and 20 nucleotides and mismatch levels from 0 to 50%, as can be seen in Fig. 1 below. Note, that this general tendency is conserved between eukaryotes and prokaryotes.

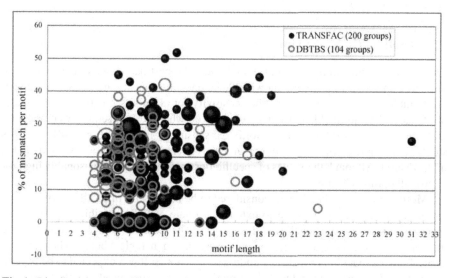

Fig. 1. Distribution of motif characteristics from various genomes. The black dots represent TRANSFAC data and the gray circles correspond to the motifs obtained from DBTBS. The size of the black dots and gray circles represents the number of motis with the same length and number of mismatches.

3.2   Selection of Important Parameters

Since there are many adjustable parameters for each program, it is impossible to test all combinations of their settings. Thus, we selected 4 parameters that severely influenced the performance of GS in an independent test. They are shown in Table 2 with their default values and the minimum and maximum values used in this study. A total of 24000 combinations of parameter values were tested.

Table 2. Parameters used for the optimization procedure. Their names, defaults, minimum and maximum values tested, and increments for each step are listed.

Parameters	Default	MIN	MAX	STEP
Number of motifs (n) [18]	10	5	100	5
Plateau period (c) [19]	20	10	200	10
Number of seeds (t) [20]	10	5	50	5
Maximum number of iterations per each seed (m) [21]	500	500	3000	500

In contrast, in our preliminary testing of MEME's parameters that do not require *a priori* knowledge of the motifs, we could not identify any crucial parameters. The only exception was that checking one modal parameter *"do not adjust motif width using multiple alignment"* slightly improved the algorithm's performance in its non-default 'nomatrim' mode, compared with the default 'trimming' mode (see 3.4). We also must mention here that MEME was used in its '-zoops' mode, in which zero or one motifs occurences are expected per sequence.

3.3   Results of Gibbs Sampler

The results obtained in the default (D) and under the best parameter values among the inspected ones (ND) for the motifs with "level one" mismatches are shown in Table 3. From Table 3 we can observe that the sensitivity for all motifs greatly increased, and in the case of 5 motifs it has become 100%. Even when 50% of sequences contain no motif occurrences, for 8 out of 13 motifs the sensitivity remains above 50%, and it is still high for some of the motifs after 60 and 90 sequences without motifs were included in the dataset. Although the values of specificities (not shown) were reasonably high for almost all optimized cases.

Motifs shown in Table 3 were considered to be at the limit of motif discovery programs' capabilities, but we tried to extend the range of GS's possibilities by experimenting on motifs, with "level two" mismatches. The results of our calculations on the datasets with these strongly corrupted motifs are shown in Table 4 below. The results in Table 4 proved that GS is capable of finding subtle motifs. It was expected to be beyond the power of the program in the default mode to extract such weak motifs, for example, 12/3, 14/4, 15/4 (motif length/number of mismatches) from the 600 nucleotides sequence, but when the appropriate parameters were applied, more than 50% of these motifs were correctly identified. A sensitivity of 28-43% was obtained for four motifs

Table 3. The sensitivity of GS for each motif with "level one" mismatches is shown. The first column shows the sensitivity in default mode (D). The second and following columns show the sensitivity in the non-default, i.e. optimized parameter mode (ND). The first two columns correspond to sensitivity averaged over datasets with a planted motif occurence in each sequence, while the columns labeled "+30", "+60" and "+90" correspond to the non-default mode sensitivity averaged over datasets which include 30, 60 and 90 extra sequences without planted motif occurences.

	D	ND	(+30)	(+60)	(+90)
8|1	28	53	12	3	0
9|1	36	82	24	8	3
10|1	100	100	98	83	56
11|2	75	86	69	40	13
12|2	68	100	100	97	46
13|3	18	82	22	3	0
14|3	35	100	100	75	35
15|3	87	100	56	38	5
16|4	30	92	21	5	0
17|4	72	100	98	100	67
18|5	70	96	56	26	7
19|5	74	96	89	51	10
20|6	23	63	15	15	3

Table 4. The sensitivity of GS for each motif with "level two" mismatches is shown. The first column shows the sensitivity in default mode (D). The second and following columns show the sensitivity in the non-default, i.e. optimized parameter mode (ND). The first two columns correspond to sensitivity averaged over datasets with a planted motif occurence in each sequence, while the columns labeled "+30", "+60" and "+90" correspond to the non-default mode sensitivity averaged over datasets which include 30, 60 and 90 extra sequences without planted motif occurences.

	D	ND	(+30)	(+60)	(+90)
8|1	0	6	0	0	0
9|1	0	3	0	0	0
10|1	0	28	8	0	0
11|2	0	3	0	0	0
12|2	2	86	3	1	0
13|3	1	6	0	0	0
14|3	0	59	10	0	0
15|3	0	60	0	0	0
16|4	0	37	15	7	0
17|4	0	39	10	5	0
18|5	0	3	0	0	0
19|5	0	43	27	3	0
20|6	0	10	0	0	0
21|7	0	71	3	0	0
22|7	24	91	50	10	0
23|8	0	70	48	7	0
30|12	0	45	28	13	0

types: 10/2, 16/5, 17/5 and 19/6, while the sensitivity of the remaining motif types was quite low.

We also investigated the sensitivity/specificity for the motifs longer than 20 nucleotides. We noticed that with the increase of motif length a higher proportion of mismatches could be tolerated. This is clearly seen when the records of 18/6 and 21/7 are compared. They have the same mismatch level of 33.3%, but very different elucidation performance.

Finally, we grouped the motifs by their lengths and the parameter values found to be the most suitable for them. These parameters showed very similar values for the motifs in the same group, and motif groups and their values are shown in Table 5 below.

Table 5. Parameters found to be the most suitable for the elucidation of correspondent groups of motifs. Motif length with the respective number of mismatches is shown as, for example, 8-9/1-2 in the first column of the table. In the last column "Sn. increase" represents the interval between minimum and maximum increase of the sensitivity for the respective motif groups.

motif group	#motifs(n)	plateau(c)	seeds(t)	iterations(m)	"Sn. increase"
default	10	20	10	500	
"short" 8-9/1-2	40-60	60-80	20	1000	3-46%
"medium" 10-15/1-4	80-100	80-100	30	1500	0-84%
"medium-long" 16-20/4-7	40-60	40	20-30	1500	3-62%
"long" 21-30/7-12	40-60	20	20-30	1500	45-71%

End-users can use Table 5 to estimate how much the elucidation performance may increase over the default setting for different motif types. We believe these settings will generally be more useful to users than the default settings.

3.4 Results of MEME

Compared to the results of GS, shown in Table 3, MEME has a much higher sensitivity for well-conserved conserved motifs or motifs with "level one" mismatches in the default setting. As shown in Table 6, the use of 'nomatrim' instead of the default 'trimming' mode showed a slight increase of 10% in several cases. After 30 sequences without motifs were added to the 30 sequences dataset with motifs, the sensitivity decreased twice or more for the most motifs.

The results in Table 7 demonstrate that our suggestion to extract subtle motifs using 'nomatrim' mode was correct. Although the general level of MEME's capability to elucidate strongly corrupted motifs is low, we can observe a bigger difference between these two modes, compared to the results demonstrated in Table 6.

4   Discussion

We aimed at improving the success rate of finding various motifs with GS and MEME. To this end, the programs were tested with various combinations of parameters. GS showed a higher dependence on parameter settings. Indeed, the sensitivity for weak motifs, which were considered to be beyond the limit of GS tolerance, almost doubled. Even in the presence of 30 sequences without motifs, the sensitivity remained higher than 50% for 8 of 13 motifs.

In our experiments with strongly corrupted motifs 3 of 13 motifs showed the sensitivity higher than 50% (12/3, 14/4, 15/4). It has been considered impossible to find such motifs in length 600 nucleotides sequence until now. Three other motifs (16/5, 17/5, 19/6) showed a sensitivity higher than 35%. We determined that for a constant

Table 6. Comparison of the sensitivity of MEME for motifs with "level one" mismatches in the default (D: 'trimming') and non-default (ND: 'nomatrim') parameter settings. "+30", "+60", "+90" corresponds to the number of sequences without the motifs in the dataset.

	D	ND	(+30)	(+60)	(+90)
8|1	60	70	2	0	0
9|1	98	99	46	17	0
10|1	100	100	68	35	2
11|2	91	95	43	11	0
12|2	98	97	45	13	0
13|3	59	58	28	8	0
14|3	93	95	21	7	0
15|3	99	99	75	37	4
16|4	89	93	0	0	0
17|4	96	97	27	12	0
18|5	89	82	0	0	0
19|5	97	96	11	4	0
20|6	66	60	0	0	0

Table 7. Comparison of the sensitivity of MEME for the motifs with "level two" mismatches in the default (D: 'trimming') and non-default (ND: 'nomatrim') parameter settings. "+30", "+60", "+90" corresponds to the number of sequences without the motif in the dataset.

	D	ND	(+30)	(+60)	(+90)
8|1	2	2	0	0	0
9|1	0	0	0	0	0
10|1	14	18	7	0	0
11|2	0	0	0	0	0
12|2	0	6	3	0	0
13|3	1	2	0	0	0
14|3	0	6	0	0	0
15|3	3	39	20	3	0
16|4	0	3	0	0	0
17|4	0	17	4	0	0
18|5	1	1	0	0	0
19|5	0	17	6	0	0
20|6	0	0	0	0	0
21|7	0	17	0	0	0
22|7	0	55	21	0	0
23|8	20	23	0	0	0
30|12	10	15	0	0	0

mismatch proportion longer motifs are easier to find. For example the 18/6 and 21/7 motifs, which both have a 33.3% mismatch level have sensitivities of 3% and 71%, respectively. For a fixed mismatch level significantly below 75% we would expect this to be the case, since the overall information content of the motif increases with motif length.

Comparing to GS, MEME could identify only a few of the subtle motifs. The best results were 15/4 and 22/7 with the correspondent sensitivities of 39% and 55%. We didn't find any great difference between default and non-default results for well-conserved motifs, but the performance become slightly better for subtle motifs when 'nomatrim' (non-default) setting is applied. For MEME we found that when increasing sequences length by 200 nucleotides, the sensitivity goes down to about of 30% (data not shown).

We found that the elucidation performance of GS strongly depends on the appropriate choice of parameters, but MEME does not show such a tendency, and generally should be applied in the default mode. We should mention that because we did not use a separate training set to optimize the parameters our estimates of sensitivity (and specificity) may be somewhat high for GS. Also it is possible that in some cases in which the programs failed to find our intended motif, they found other patterns of similar or greater strength. We recognize this interesting possibility but have not systematically investigated it at this point. As for computational resources, we have not reported the specific execution times here but both programs were generally sufficiently fast (seconds to a few minutes) with GS generally being faster than MEME.

In conclusion, we have performed a parametric analysis with the popular MEME and GS programs. In particular, we show that appropriate settings can greatly improve the performance of GS over its default settings. We classified motifs in several groups based on their length and assigned them parameters values suitable for effective elucidation. In addition to providing hints to algorithm designers, our results should be immediately helpful to end-users.

Acknowledgements

We would like to thank Tim Bailey for helpful comments.

References

1. Eisen, M., Spellman, P., Brown, P., Botstein, D.: Cluster analysis and display of genome-wide expression patterns. PNAS, Vol. 95. (1998) 14863-14868
2. Stormo, G.: DNA binding sites: representation and discovery. Bioinformatics, Vol. 16. (2000) 16-23
3. Stormo, G., and Hartzell, G.: Identifying protein-binding sites from unaligned DNA fragments. PNAS, Vol. 86. (1989) 1183-1187
4. Lawrence C., and Reilly, A.: An Expectation Maximization (EM) Algorithm for the Identification and Characterization of Common Sites in Unaligned Biopolymer Sequences. Proteins, Vol. 7. (1990) 41-51
5. Lawrence, C., Altschul, S., Boguski, M., Lui, J., Neuwald, A., Wootton, J.: Detecting subtle sequence signals: A Gibbs sampling strategy for multiple alignment. Science, Vol. 262. (1993) 208-214
6. Bailey, T., and Elkan, Ch.: Unsupervised learning of multiple motifs in biopolymers. Machine Learning, Vol. 21. (1995) 51-80
7. Frith, M., Hansen, U., Spouge, J., Weng, Z.: Finding functional sequence elements by multiple local alignment. Nucl.Acid Res., Vol. 32 (2004) 189-200
8. Hertz, G., and Stormo, G.: Identifying DNA and protein patterns with statistically significant alignments of multiple sequences. Bioinformatics, Vol. 15. (1999) 563-577
9. Horton, P.: Tsukuba BB: A Branch and Bound Algorithm for Local Multiple Alignment of DNA and Protein Sequences. Journal of Computational Biology, Vol. 8. (2001) 249-282
10. Sinha, S., and Tompa, M.: Discovery of novel transcription factor binding sites by statistical over-represenatation. Nucleic Acids Res., Vol. 30. (2002) 5549-5560
11. Keich, U., and Pevzner, P.: Subtle motifs: defining the limits of motif finding algorithms. Bioinformatics, Vol.18. (2002) 1382-1390

12. Yada, T., Totoki, Y., Ishikawa, M., Asai, Nakai, K.: Automatic extraction of motifs represented in the hidden Markov model from a number of DNA sequences. Bioinformatics, Vol.14 (1998) 317-325

13. Pevzner, P., and Sze, S.: Combinatorial approaches to finding subtle signals in DNA sequences. In Proceedings of the 5th International Conference on Intelligent Systems for Molecular Biology (ISMB), (2000) 269-278

14. Sze, S., Gelfand, M., Pevzner, P.: Finding weak motifs in DNA sequences. In Proceedings of the Pacific Symposium of Biocomputing (PSB), Vol. 7. (2002) 235-246

15. Poluliakh, N., Takagi, T., Nakai, K.: MELINA: motif extraction from the promoter regions of *co*-regulated genes. Bioinformatics.Vol. 19 (2003) 423-424

16. Wingender, E., Chen, X., Hehl, R., Karas, H., Liebich, I., Matys, V., Meinhardt, T., Pruss, M., Reuter, I., Schacherer, F.: TRANSFAC: an integrated system for gene expression regulation. Nucleic Acids Res., Vol. 28. (2000) 316-319

17. Makita, Y., Nakao, M., Ogasawara, N., and Nakai, K.: DBTBS: Database of transcriptional regulation in Bacillus subtilis and its contribution to comparative genomics, Nucleic Acids Res, 32. (2004) 75-77

18. *n* - An initial "guesstimate" of the total number of motifs in all of the sequences

19. *c* - The *Plateau* period is the number of iterations between successive local maxima. For each seed, the program samples until *Plateau* period iterations are performed without an increase in the MAP value. (MAP value (maximum *a posteriori*) is a measure of the statistical significance of the motifs alignment comparing to a "*null*" alignment.)

20. *t* - The number of times the sampler restarts with different seeds.

21. *m* - Maximum number of iterations per seed.

Inferring Cis-region Hierarchies from Patterns in Time-Course Gene Expression Data

Vladimir Filkov[1] and Nameeta Shah[1,2]

[1] CS Dept., UC Davis
One Shields Avenue, Davis, CA 95616
filkov@cs.ucdavis.edu
[2] CIPIC, UC Davis
One Shields Avenue, Davis, CA 95616
shahn@cs.ucdavis.edu

Abstract. Resolving the co-regulation relationships between genes is a major step toward understanding the underlying topology and dynamics of gene networks. Although co-expression of genes does not directly imply their co-regulation, model-based approaches coupled with the availability of large-scale gene expression data can help associate expression patterns with features in their cis-regions. Inspired by studies of transcriptional regulation in sea-urchin, here we report on preliminary validation of the following simple model for transcriptional regulation in yeast: the same Cis-Regulatory Modules (CRMs) in the cis-regions of different genes give rise to very similar functional events in the time-course expression profiles of those genes. We use a modified version of a prior algorithm for decomposing time-course gene expression patterns into functional events. To capture and reason about shared CRMs we introduce an order relationship, or a *Regulation Hierarchy* on the genes. When tested on actual time-course gene expression data of yeast preliminary results indicate 50% - 71% matches, of high confidence, between our derived and known cis-region regulation hierarchies. This hierarchy structure yields practical predictions when used with other type of genomic data, e.g. location of TF-DNA interactions.

1 Introduction

Gene expression is regulated during transcription by combinations of trans-factors (TFs) that bind to corresponding sites in the genes' cis-regions. The differential expression of any gene under different experimental conditions is due to the particular regimen (i.e. abundances) of those TFs. Exactly how the cis-regions process the input protein concentration signals is a key question in functional genomics. Methods for grouping genes by similarity of expression profiles across multiple experiments have been partially successful in identifying functionally related genes [1]. But since co-expression does not imply co-regulation in general such methods have been limited to identification of gross functional features and categories.

The next step is to incorporate known biological facts into computational models of regulation. Such model-based approaches coupled with the availability of large-scale gene expression data can help associate expression patterns with features in their cis-

E. Eskin, C. Workman (Eds.): RECOMB 2004 Ws on Regulatory Genomics, LNBI 3318, pp. 88–97, 2005.

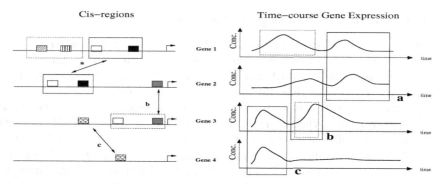

Cis–regions   Time–course Gene Expression

Fig. 1. On the left are given cis-regions for four genes and the cis-elements in them. On the right hand side are the corresponding gene expression signals. **a, b,** and **c** are the modules of cis-elements and their effects on the expression. The dashed boxes indicate expression events for which there might be multiple causes.

regions, and thus elucidate co-regulation. The choice of a model then becomes an important issue, especially since there are very few good qualitative models out there and no complete quantitative model of general transcriptional regulation, to the best of our knowledge.

Perhaps one of the best qualitative descriptions of the inner-workings of cis-regions (CRs) has come out of the work of Eric Davidson [2] and colleagues on sea urchin. They have demonstrated that single or groups of TF binding sites in cis-regions behave as functional units of the regulatory systems, called *Cis Regulatory Modules, CRMs.* Davidson and colleagues found out, using many genes in the Sea Urchin organism [3], that the function of their cis-regions, i.e. gene expression signals, can be decomposed into simpler functions of their sub-regions, down to the functions of the individual CRMs. The CRMs, then, can be thought of as the building blocks or transcriptional regulation, each having identifiable functional manifestation, i.e. gene expression pattern. Thus the CRM identification is extremely important.

Ideally, one would like to formalize their model and use it to resolve the complexity of cis-regions from large-scale functional genomics data, like gene expression microarray data for example. As a prelude to such a formalization, in this paper we sought to test whether the modularity parallel between cis-regions and gene expression can be detected from available large-scale functional genomics data. Our working hypothesis is that *shared events, or sub-signals, of gene expression signals are due to modules of shared binding sites in the cis-regions of the genes.* Fig. 1 gives an illustration. There, the sub-signals **a, b** and **c** are consequences of the actions of the corresponding modules in the cis-regions.

Before proceeding, we need a somewhat formalized notion of cis-modularity. Drawing from the discussion and illustration above we come up with the following rules:

1. CRMs are the smallest sets of binding sites which have distinguishable function;
2. CRMs functions are manifested as sub-signal of the genes' expression signals; and
3. Two or more modules on the same cis-region are responsible for expression regulation at different times or places in the organism; otherwise they would be considered a single module.

In this work, to test our working hypothesis above,

- We propose identifying expression events by decomposing gene expression signals into *Putative Elementary Expression Events (PEEEs)* using a modification of our previous approach [4], which was developed for elucidating regulatory relationships between pairs of genes from time-course expression data.
- We introduce the *Regulation Hierarchy* structure as a representation of co-regulation between genes. The Regulation Hierarchy is a directed acyclic graph, in which genes are partially ordered based on shared cis-modules. In such a graph any two nodes with a common ancestor are co-regulated. Such a hierarchy graph is useful independently as a structure for study of functional elements of gene regulation. As upper and lower bound approximations of the Regulation Hierarchy we define two other hierarchies, the *Expression Hierarchy* and the *Transcription Factor Hierarchy* which can be obtained from existing data sets.
- We compare the Expression and Transcription Factor Hierarchies obtained from two publicly available data sets: a time-series, whole-genome, expression data of yeast [5], and a genome-wide location data of TF-DNA binding in yeast [6]. Our results suggest strong correlation between sub-elements of gene expression curves and cis-modules of binding sites: we observed 50% - 71% of matching directed edges between the two herarchies, compared to expected (between $1/6$ and $1/8$ of that). By combining the predicted hierarchies we were able to discern basic expression signals and attribute some to well known TF modules.

The paper is organized as follows. Next we talk about related work on effects of modularity of the cis-regions on gene expression, and decomposition of expression signals into basic curves. In Section 2.1 we review and expand a previous method for identifying elementary expression events. The regulation hierarchies are defined in Sect. 2.2, and we show how to construct them from real data in Sect. 3. We report the results of our preliminary studies in Section 4. In the last section we summarize the findings and describe our current and future directions in both expanding the model and utilizing the hierarchy graphs in different ways.

1.1   Co-regulation and Co-expression

Our current work is novel in that it proposes a model for co-regulation based on attributing identifiable events in expression signals to cis-modules. We also describe an original structure, the Regulation Hierarchy.

Differentiating between co-expressed and co-regulated genes is important in particular for gene network inference. In previous work Pilpel et al. proposed [7] and later improved [8] methods to identify clusters of genes which are co-regulated and co-expressed at the same time. They achieved this by scoring co-expression for genes which share overrepresented elements in the upstream regions. Although our goal is seemingly the same, since we are also attempting to resolve co-regulation from co-expression, here we are interested in resolving co-regulation from expression data, based on the shared events model. We don't aim to resolve actual binding sites in this paper; instead we are after the co-regulation hierarchy, for which we only use time-course expression data. We discuss later some future uses for the Regulation Hierarchy.

A few studies recently have focused on identifying modules of genes by considering variety of available data: gene expression, sequence, and TF-DNA location. The working definition for a module in them varies between a group of strongly co-expressed genes in a subset of experiments to a group of genes co-regulated by the same factors and sharing a function [9]. In both extremes though the definition of a module is somewhat fuzzy as genes can be taken in or out of it while the module doesn't change. Our definition of a cis-module, a variant of that of Davidson [2], is a group of transcription factors that has an indivisible functional effect on transcription; in other words there is a sense of minimality or atomicity to it. Time-course expression data cannot predict the actual content of a cis-module; other data is needed for that (gene knockout expression data can also be used).

The small number of different patterns evident in time-course gene expression data, especially the cycling genes set by Spellman et al. [5], has motivated several studies into evaluating the possibility of decomposing the expression signals into a combination of a few basic signals. In particular the study by Holter et al. [10] identifies a small number of characteristic modes in microarray time-series data [5], as discovered by Singular Value Decomposition. Such studies although informative about the range of the transcriptional signals under specific conditions, and arguably successful in correlating functional gene categories with specific modes of regulation, do not address the issue of co-regulation.

2   Elementary Expression Events and Regulation Hierarchies

In this section we formalize the notion of shared features in the genes expression signals by introducing *Putative Elementary Expression Events (PEEEs)*. The intuition, and a simplification of our working hypothesis, is that shared PEEEs correspond to shared CRMs. To reason about shared PEEEs and their relationship to shared CRMs in the corresponding cis-regions we introduce a graph theoretic structure, the *Regulation Hierarchy*, and a few derivatives.

2.1   Expression Events

PEEEs are functionally relevant parts of the expression signals and, we postulate, are functional effects of CRMs. For our purposes, these are parts of the expression signals that either increase or decrease. They are identified using a modified version of the edge detection algorithm by Filkov et al. [4].

There, events were defined as biologically meaningful changes in expression with time. In the ideal case, with no fluctuations in the signals, events would correspond to monotonically increasing or decreasing smooth curves between local optima. Because large-scale gene expression data is far from ideal, signals are smoothed out as follows. Starting from the initial time point, and proceeding to the right iteratively, over the rest of the time points, the events are identified, grown, and possibly merged, so long as the expression change is in the same direction (i.e. increase or decrease) as the rest of the event, with tolerances for default and random fluctuations in expression levels, as well as with a biologically significant cap on the maximum length of an event. The original method uses one neighbor on both sides of time points to label them local minima, local maxima and in-between. But edges can be missed that way because of noise in the data. We improve on this by using two neighbors on both sides of a local optimum to label the points more accurately.

The result is a list of putative events, or PEEEs, for each gene. Each event is a run of points that either increases or decreases in expression. We showed previously [4] that these lists of events can be used to identify gene regulatory relationships between genes with greater fidelity than co-expression. In addition, the putative events lists for pairs of genes can be aligned to discover any shared events.

2.2   Regulation Hierarchy Graphs

In our model of transcriptional regulation a gene's expression is completely determined by the modules in its cis-region. The idea behind the Regulation Hierarchy is to build a structure that captures the shared regulation information between genes. The Regulation Hierarchy is meant to be an invariant view of regulation from both the sequence and gene expression, and a representation of both.

The *Regulation Hierarchy (RH)* is defined as a directed graph, $G_r = (V, E_r)$ over the genes in an organism, $V = \{g_1, g_2, \ldots, g_n\}$, where there is an edge between two nodes if the set of CRMs regulating one gene is a subset of the set of the CRMs regulating the other, and the direction of the edge is from the smaller toward the larger set of regulators. That is, if $Mod(x)$ is the set of modules regulating node x, then for every pair of genes (nodes) i and j, $(i, j) \in E_r$ if $Mod(i) \subseteq Mod(j)$. If i and j share CRMs but none dominates the other, then neither $(i, j) \in E_r$ nor $(j, i) \in E_r$. For example, there is only one such relationship between the genes in Fig. 1, and that is $Gene4 \leq Gene3$. The rest of the gene pairs don't have order relationships although they share regulators (and sub-signals).

We define the following two additional hierarchy graphs, which, in contrast to the regulation hierarchy, can be obtained from existing data. First is the *TF hierarchy (TFH)*, defined as $G_{tf} = (V, E_{tf})$, where if $Tf(x)$ is the set of transcription factors that can bind to the cis-region of gene x then $(i, j) \in E_{tf}$ if $Tf(i) \subseteq Tf(j)$. The second hierarchy is the *Expression Hierarchy (EH)*, defined as $G_e = (V, E_e)$, where if $Peee(x)$ is the set of PEEEs present in the expression signal of gene x then $(i, j) \in E_e$ if $Peee(i) \subseteq Peee(j)$.

Properties of the regulation hierarchies. The Expression and Transcription Factor Hierarchies are, in a way, an upper and lower bound (respectively) on the edges in the Regulation Hierarchy, because $E_e \subseteq E_r \subseteq E_{tf}$. We see that as follows.

First of all, $E_r \subseteq E_{tf}$. Namely, if $(i, j) \in E_r$ then $Mod(i) \subseteq Mod(j)$, but that implies $Tf(i) \subseteq Tf(j)$ since modules are just sets of TFs. Hence, $(i, j) \in E_{tf}$. Note that these two hierarchies will not be equal in reality because there can be TFs binding to a cis-region without having a functional effect on that gene's expression.

Second, $E_e \subseteq E_r$, because as we assumed in the CRM rules before, each modules has an identifiable functional sub-component of the expression signal, and the modules' functions are non-overlapping. Thus each PEEE corresponds to a module. Note that these two hierarchies will not be equal in reality because not all modules' functions are identifiable from existing data.

With ideal (but not necessarily complete) data, these three hierarchies would be directed, and transitively closed graphs, where if $(i, j) \in E$ and $(j, k) \in E$ then $(i, k) \in E$. They would also be acyclic except for the trivial cycles which will happen between two genes sharing exactly the same regulators.

Utility of the regulation hierarchies. From the RH one can readily answer if two genes are co-regulated by looking up if they have the same ancestor. Also, with the RH and the TFH one can explore TF modules, whereas from the RH and EH the basic expression signals corresponding to modules can be found.

In addition, RH can be a powerful tool for building regulatory networks. Namely, the RH establishes classes of co-regulated genes–information that can help bound the in-degrees of nodes during inference.

In this paper we show how to obtain the Expression Hierarchy and the TF hierarchy, and explore how well they coincide. Ideally, $E_e \subseteq E_{tf}$. We use both to illustrate how one can identify TF modules.

3   Constructing the Hierarchies from Existing Data

We used two separate data sets of yeast. The first is a time-course, genome-wide, gene expression data, known as the Cell Cycling Genes data, from Spellman et al. [5] Although somewhat dated, we used this data set because it is still one of the best time-course expression data sets available, mostly because of the long length of the series (i.e. number of measurements is large) as well as the sampling times (they are small enough to capture the cell-cycling processes in yeast). The data set consists of four separate time-series measurements of expression for each gene, totaling 76 measurements, for about 6200 genes of yeast. We imputed the missing values using KNNimpute [11]. We concatenated all the measurements and obtained a 76 dimensional, real-valued, vector for each gene.

The second data set is the TF-DNA data by Lee et al. [6]. The set consists of 6200×106 p-values indicating the confidence of binding for each of 106 TFs to all 6200 genes of yeast. By selecting a confidence value for each gene one obtains a TF profile of binding (i.e. a list of TFs that bind to the closest intergenic region to that gene). We used $p = 0.01$ as the threshold.

The Cell-Cycling Genes data set does not have too many features, i.e. the expression signals do not have many degrees of freedom as the conditions to which the genes were exposed in that experiment were not diverse. Thus one needs to lower the dimensionality of the expression matrix, since over 6000 signals present an overkill and will result in a large number of spurious events identified. So, we clustered the data into a smaller number of clusters which should all be sufficiently different and offer variety of sub-signals. The genes' expression vectors were clustered using average-link hierarchical clustering with the Pearson's correlation as the distance measure. The clustering goal was indistiguishability of curves within clusters under visual observation. The resulting 87 clusters are the nodes in the hierarchy graphs.

For the Expression Hierarchy we created an average expression profile for each cluster by averaging the expression vectors from the Cell Cycling genes data. We ran our modified edge detection algorithm to detect events (see Section 2.1). An event profile was created for each gene, consisting of runs of points labeled as increasing or decreasing. To identify common events pairs of profiles were overlapped and the events matching in location and direction were counted. For the TF Hierarchy we created a regulation profile for each gene using the TF-DNA data. Then we calculated the overlap score between two clusters, from the expression data clustering, as defined above.

3.1 Constructing EH and TFH

Real data of course is noisy. Thus, we had to allow for some fuzziness in the regulation hierarchies. In addition, the data sets with which we worked were clusters of gene expression signals as opposed to individual signals. Thus each cluster contains a number of co-expressed genes, whose curves are to us indistinguishable. Here we describe how we derived the expression and TF hierarchies from noisy clusters of genes.

Expression hierarchy. The data used for the EH is time-course expression data (see above) from which PEEEs have been identified for each gene. An edge was created between two nodes based on the overlap score of their PEEE lists. The average of the expression signals in a cluster was the representative signal for that cluster. Then the PEEE list for that average was the PEEE list for that node.

We define the overlap score for edge (i, j) by using a combination of the following two scores:

(i) $S_{ij} = |Peee(i) \cap Peee(j)|$, i.e number of common PEEEs present in the event sets of both nodes;
(ii) $S_{i-j} = |Peee(i)| - |Peee(i) \cap Peee(j)|$, i.e. number of PEEEs present in node i but not in j.

Then,

$$(i, j) \in E_e \text{if } S_{ij} > \overline{S_{ij}} + Z_e \sigma \text{ and } \frac{S_{i-j}}{S_{ij}} < 0.3 \qquad (1)$$

We considered an edge present if the overlap was Z_e standard deviations more than average. This z-score served as our threshold for the edges in E_e. To ensure the containment relationship but allow for noisy data we added the constraint that S_{i-j} be less than 30% of the overlap. In other words, for an edge (i, j) we allow for some PEEEs to be in i but not in j. The 30% we determined to be a well balanced cap on such events.

TF hierarchy. The data used for the TFH was TF-DNA interaction location data (see above). An edge (i, j) in the TFH graph was established by carefully evaluating the overlapping and non-overlapping sets of TFs between nodes i and j. As the nodes are clusters of genes, we counted the overlap between pairs within and between clusters.

An edge (i, j) is defined by using a combination of following three scores:

(i) A, inter-cluster overlap of TFs. The score is obtained by counting the number of common TFs for all pairs formed by genes in cluster i and cluster j.
(ii) B, intra-cluster overlap of TFs. The score is obtained by counting the number of common TFs for all pairs formed by genes in cluster i.

(iii) C, intra-cluster overlap of TFs. The score is obtained by counting the number of common TFs for all pairs formed by genes in cluster j.

Then,

$$(i, j) \in E_{tf} \text{ if } A > \overline{A} + Z_{tf}\sigma \text{ and } B < C \tag{2}$$

We considered an edge present if the TF overlap was Z_{tf} standard deviations more than average. To ensure the containment relationship we added the constraint that B be less than C.

4   Preliminary Results

To test our model of regulation we compared the two resulting hierarchies, the EH and TFH. In ideal conditions, $E_e \subseteq E_{tf}$, i.e. all the edges in EH should be in TFH.

As a part of our preliminary studies we built several different EH and TFH, for different values of the z-scores Z_e and Z_{tf}. We also generated random graphs on the 87 nodes by permuting the expression data and running the algorithms to identify PEEEs and score them on this permuted data. The results for $Z_{ef} > 0.5$ and $Z_{ef} > 1$ are shown in Fig. 2.

Several things are evident from the figure. First of all the edges in the Expression Hierarchy correspond very well to the edges in the TF Hierarchy: with increasing Z_e we get up to 71% matches. The second observation is that the results are significant: the random graphs have many fewer edges (down to about 10%) that match the TFH. So we did in fact get most of the edges from the EH in TFH.

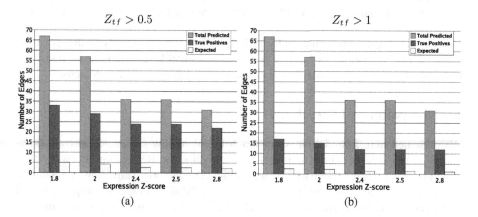

Fig. 2. Comparing the inferred Expression Hierarchy to a known TF Hierarchy, at two different thresholds of TF factor overlap $Z_{tf} > 0.5$ (560 edges) and $Z_tf > 1$ (290 edges). The total number of edges in E_e, for 5 varying thresholds of overlap, together with the true positives and the expected (random) matches is given. The correct predictions increase from 48% and 71%, with the expected number decreasing to 1/6 of that (for $Z_{tf} > 0.5$). In (b) although the statistical significance of the results is better than in (a) the sensitivity is lowered.

The number of edges in EH is about 10% to 20% of those in TFH. This was to be expected as the genes were only exposed to a few different conditions during the Cell Cycling Genes experiment, so the actual range of expression signals captured only a small number of modules' effects.

The resulting clusters and hierarchy graphs can be obtained from our Web site: graphics.cs.ucdavis.edu/~nyshah/Regulation. Here we omit them for space considerations and report on some initial observations. First of all the hierarchy of the yeast genome is shallow with the longest paths being of length at most 5. This is well in agreement with other studies [12].

Next we examined co-regulated nodes in the partially ordered graph by comparing the TFs from the TF-DNA location data set. Note that each node is a cluster of genes. The following are modules of TFs that were common to at least three genes between co-regulated nodes. Together with the genes we give some annotation either from SGD [13] or individual references. Some of the TFs in the modules are known to act together while the others are not. More TF modules are available at our Web site.

Cluster 4 (12 genes) 4 regulated by FKH2, MCM1 and NDD1. MCM1 is a known yeast cell cycle regulator during the M and M/G1 phases. FKH2 is involved in the regulation of the SIC1 cluster, whose member genes are expressed in the M/G1 phase of the yeast cell cycle, and are involved in mitotic exit [14]. NDD1 is a high-dosage suppressor of cdc28-IN, essential for expression of a subset of late S phase specific genes in yeast [15].

Cluster 6 (12 genes) 4 regulated by all of GAL4, GAT3, RGM1, YAP5. GAL4 is a well known transcription factor for the GAL structural genes, which encode galactose metabolic proteins. GAT3 (YLR013w) is a protein encoding GATA-family zinc finger motifs, known transcription factors [16]. RGM1 is a putative transcriptional repressor with proline-rich zinc fingers. YAP5 is a bZIP protein and a known transcription factor.

Cluster 13 (25 genes) 6 regulated by MBP1, SWI6. SWI6 is a transcription cofactor, forms complexes with DNA-binding proteins Swi4p and Mbp1p to regulate transcription at the G1/S transition. MBP1 is a cell-cycle regulating transcription factor.

5   Discussion and Directions

We presented here a model-based approach to elucidating co-regulation from time-course gene expression measurements. We introduce the Regulation Hierarchy as a structure that usefully summarizes transcriptional regulation, show how it relates to two more practical structures, the expression and TF hierarchies, and approximate it using gene expression data. Using publicly available data on gene expression and TF-DNA binding in yeast we were able to get encouraging results supporting the utility of the Regulation Hierarchy, and its derivation from expression data. We demonstrate one particular use for the RH by combining it with the TF-DNA data and identifying TF regulatory modules.

Again these are preliminary studies, and there are many things on which we need to improve. Our method for identifying PEEE is ad hoc and dated; better methods from time-series analysis will likely yield better PEEEs. The overlap scores can also be improved upon by doing alignment and maximal overlaps for example.

The biggest goal in front of us is building the regulation hierarchies from static expression data, of which there is thousands of available sets for yeast. If that is possible, the resulting expression hierarchy would have many more edges, as the genes would have been exposed to significantly more conditions than the ones in the data we used.

The regulation hierarchies are not ends in themselves but stepping stones toward identifying interactions between genes and gene products on a large-scale. In particular, they can be used jointly with gene expression data to limit the in-degree of nodes during network inference, which can speed up the process significantly.

References

1. Eisen, M., Spellman, P., Brown, P., Botstein, D.: Cluster analysis and display of genome-wide expression patterns. Proceedings of the National Academy of Science **85** (1998) 14863–14868

2. Davidson, E.: Genomic Regulatory Systems. Academic Press (2001)

3. Davidson, E., et al.: A genomic regulatory network for development. Science **295** (2002) 1669–1678

4. Filkov, V., et al.: Analysis techniques for microarray time-series data. Journal of Computational Biology **9** (2002) 317–330

5. Spellman, P., Sherlock, G., Zhang, M., Iyer, V., Anders, K., Eisen, M., Brown, P., Botstein, D., Futcher, B.: Comprehensive identification of cell cycle-regulated genes of the yeast saccharomyces cerevisiae by microarray hybridization. Molecular Biology of the Cell **9** (1998) 3273–3297

6. Lee, T., et al.: Transcriptional regulatory networks in *saccharomyces cerevisiae*. Science **298** (2002) 799–804

7. Pilpel, Y., Sudarsanam, P., Church, G.: Identifying regulatory networks by combinatorial analysis of promoter elements. Nature Genet. **29** (2001) 153–159

8. Lapidot, M., Pilpel, Y.: Comprehensive quantitative analyses of the effects of promoter sequence elements on mrna transcription. Nucleic Acids Research **31** (2003) 3824–3828

9. Segal, E., Shapira, M., Regev, A., Pe'er, D., Botstein, D., Koller, D., Friedman, N.: Module networks: Identifying regulatory modules and their condition specific regulators from gene expression data. Nature Genetics **34** (2003) 166–76

10. Holter, N., et al.: Fundamental patterns underlying gene expression profiles:simplicity from complexity. PNAS **97** (2000) 8409–8414

11. Troyanskaya, O., Cantor, M., Sherlock, G., Brown, P., Hastie, T., Tibshirani, R., Botstein, D., Altman, R.: Missing value estimation methods for dna microarrays. Bioinformatics **17** (2001) 520–525

12. Alon, U. Internationl Conference on Systems Biology (2003) Invited Talk.

13. Dolinski, K., et al.: Saccharomyces genome database (2004) http://www.yeastgenome.org/.

14. Zhu, G., et al.: Two-yeast forkhead genes regulate the cell-cycle and pseudohyphal growth. Nature **406** (2000) 90–94

15. Loy, B., et al.: Ndd1, a high-dosage suppressor of cdc28-in, in sacc. cerevisiae. Mol. Cell. Biol. (1999)

16. Cox, K., Pinchak, A., Cooper, T.: Genome-wide transcriptional analysis in s. cerevisiae by mini-array membrane hybridization. Yeast **15** (1999) 703–713

Modeling and Analysis of Heterogeneous Regulation in Biological Networks

Irit Gat-Viks*, Amos Tanay*, and Ron Shamir

School of Computer Science, Tel-Aviv University, Tel-Aviv 69978, Israel
{iritg,amos,rshamir}@tau.ac.il

Abstract. In this study we propose a novel model for the representation of biological networks and provide algorithms for learning model parameters from experimental data. Our approach is to build an initial model based on extant biological knowledge, and refine it to increase the consistency between model predictions and experimental data. Our model encompasses networks which contain heterogeneous biological entities (mRNA, proteins, metabolites) and aims to capture diverse regulatory circuitry on several levels (metabolism, transcription, translation, post-translation and feedback loops among them).

Algorithmically, the study raises two basic questions: How to use the model for predictions and inference of hidden variables states, and how to extend and rectify model components. We show that these problems are hard in the biologically relevant case where the network contains cycles. We provide a prediction methodology in the presence of cycles and a polynomial time, constant factor approximation for learning the regulation of a single entity. A key feature of our approach is the ability to utilize both high throughput experimental data which measure many model entities in a single experiment, as well as specific experimental measurements of few entities or even a single one. In particular, we use together gene expression, growth phenotypes, and proteomics data.

We tested our strategy on the lysine biosynthesis pathway in yeast. We constructed a model of over 150 variables based on extensive literature survey, and evaluated it with diverse experimental data. We used our learning algorithms to propose novel regulatory hypotheses in several cases where the literature-based model was inconsistent with the experiments. We showed that our approach has better accuracy than extant methods of learning regulation.

1 Introduction

Biological systems employ heterogeneous regulatory mechanisms that are frequently intertwined. For example, the rates of metabolic reactions are strongly coupled to the concentrations of their catalyzing enzymes, which are themselves subject to complex genetic regulation. Such regulation is in turn frequently affected by metabolite concentrations. Metabolite-mRNA-enzyme-metabolite feedback loops have a central role in many biological systems and exemplify the importance of an integrative approach to the modeling and learning of regulation.

---

* These authors contributed equally to this work.

E. Eskin, C. Workman (Eds.): RECOMB 2004 Ws on Regulatory Genomics, LNBI 3318, pp. 98–113, 2005.
© Springer-Verlag Berlin Heidelberg 2005

In this work we study steady state behavior of biological systems that are stimulated by changes in the environment (e.g., lack of nutrients) or by internal perturbations (e.g., gene knockouts). Our model of the system contains variables of several types, representing diverse biological factors such as mRNAs, proteins and metabolites. Interactions among biological factors are formalized as regulation functions which may involve several types of variables and have complex combinatorial logic. Our model combines metabolic pathways (cascades of metabolite variables), genetic regulatory circuits (sub-networks of mRNAs and transcription factors protein variables), protein networks (cascades of post-translational interactions among protein variables), and the relations among them (metabolites may regulate transcription, enzymes may regulate metabolic reactions). We show how such models can be built from the literature and develop computational techniques for their analysis and refinement based on a collection of heterogeneous high-throughput experiments. We develop algorithms to learn novel regulation functions in lieu of ones that manifest inconsistency with the experiments.

Most current approaches to the computational analysis of biological regulation focus on transcriptional control. Both discrete (e.g., [3]) and probabilistic methods (e.g., [9]) use gene expression data and attempt to learn a regulatory structure among genes and to create a predictive model that fits the data. The computational models used in these studies involve numerous simplifying assumptions on the nature of genetic regulation. Among the more problematic of these simplifications are a) the use of mRNA levels to model the activity of transcription factor proteins, b) the lack of consideration for the state of the medium in which the experiment was done and c) the assumption of acyclic regulation structure that prevents the adequate modeling of feedback loops. As a consequence of these limitations, simple genetic networks tools are rarely used in practical biological settings. A more fruitful approach for learning regulation involves the coarser notion of regulatory modules, with [14] or without [1, 17] explicit learning of regulatory functions that define them. Module-based methods are relatively robust to noise and in some cases can tolerate the gross simplification described above. However, models generated by these methods are coarse and limited in their level of detail.

Our study aims to overcome some of the limitations of prior art by taking an approach that is innovative in combining several key aspects:

• We model a variety of variables types, extending beyond gene network studies, that focus on mRNA, and metabolic pathways methods, that focus on metabolites. Consequently, our model can express the environmental conditions and the effects of translation regulation and post translational modifications.

• Our approach allows handling feedback loops as part of the inference and learning process. This is crucial for adequate joint modeling of metabolic reactions and genetic regulation.

• We build an initial model based on prior knowledge, and then aim to improve (expand) this model based on experimental data. A similar approach was employed in [16] for transcription regulation only. We show that formal modeling of the prior knowledge allows the interpretation of high throughput experiments on a new level of detail.

• Our algorithms learn new transcription regulation functions by analyzing together gene expression, protein expression and growth phenotypes data.

Our methodologies and ideas were implemented in a new software tool called MetaReg. It facilitates evaluation of a model versus diverse experimental data, detection of variables that manifest inconsistencies between the model and the data, and learning optimized regulation functions for such variables. We used MetaReg to study the pathway of lysine biosynthesis in yeast. We performed an extensive literature survey and organized the knowledge on the pathway into a model consisting of about 150 variables. In the process of model construction, we reviewed the results of many low throughput experiments and included in the model the most plausible regulation function of each variable. We assessed the model versus a heterogeneous collection of experimental results, consisting of gene expression, protein expression and phenotype growth sensitivity profiles. In general, the model agreed well with the observations, confirming the effectiveness of our strategy. In several important cases, however, inconsistencies between measurements and model predictions indicated gaps in the current biological understanding of the system. Using our learning algorithm we generated novel regulation hypotheses that explain some of these gaps. We also showed that our method attains improved accuracy in comparison to extant network learning methods.

The paper is organized as follows. In Section 2 we introduce the model and define some notation. In Section 3 we show how to take feedback loops into account and how to use the model to infer the system state given an environmental stimulation. In Section 4 we introduce our mathematical formulation of experimental data and model scoring scheme and in Section 5 we develop optimization algorithms for the learning of regulation functions. Section 6 presents our results on the lysine pathway and its regulation.

2   The Model

We first define a formal model for biological networks. A *model* M is a set U of *variables*, a set $S = \{1, \ldots, k\}$ of discrete *states* that the variables may attain, and a set of *regulation functions* $f_v : S^{|N(v)|} \to S$ for each $v \in U$. f_v defines the state of a regulated variable v (called a *regulatee*) as a function of the states of its *regulator* variables $N(v) = \{r_v^1, \ldots, r_v^{d_v}\}$. We define the set of *stimulators* U_I to include all variables with zero indegree. The *model graph* of M is the digraph $G_M = (U, A)$ representing the direct dependencies among variables, i.e., $(u, v) \in A$ iff $u \in N(v)$. For convenience we assume throughout that regulation functions can be computed in constant time.

A *model state* s is an assignment of states to each of the variables in the model, $s : U \to S$. A model *stimulation* is an assignment of states to all the model stimulators, $q : U_I \to S$.

In this paper we shall use the model logic primarily for the determination of modes. For a model M and state s, we say that *s agrees with M on* v if $f_v(s(r_v^1), \ldots, s(r_v^{d_v})) = s(v)$. We call a model state s of M a *mode* if s agrees with M on every $v \in U \setminus U_I$. A mode is thus a steady state of the system. States representing non-steady state behavior of the system, which may be adequate for the representation of temporal processes, are outside the scope in this work. Since our biological models represent a combination of diverse regulation mechanisms, operating in different time scales (metabolic reactions are orders of magnitude faster than transcription regulation), a realistic temporal model is a considerable challenge that should be carefully dealt with in future work. The steady state assumption is in wide use (e.g., [3, 9]) and was proved flexible enough in our empirical studies. Figure 1 illustrates a simple model and its modes.

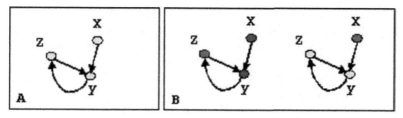

Fig. 1. A simple model. The model includes one stimulator X, regulating a positive feedback loop of two variables Y and Z. We assume a binary state space (on-dark, off-light). f_z is the identity function and $f_y = s(x) \; AND \; s(z)$. When the stimulator state is off (A), a unique mode exists. If the stimulator state is on (B), two different modes are possible, one in which the cycle is on and the other in which the cycle is off.

We now describe the biological semantics of a model. V includes four *types* of variables: (a) mRNAs (b) active proteins that serve as enzymes or regulators (c) internal metabolites, which represent the metabolite derivatives in the pathway under study (d) external metabolites, which represent different environmental conditions and specify the nutritional concentrations in the medium. The external metabolites are assumed to be determined by the experimenter, and their level is unaffected by other variables in the model, so they will serve as part of our stimulator set. The levels of the mRNAs, proteins and internal metabolites are controlled by other variables via regulation functions that manifest transcriptional, translational, post-translational and metabolic control mechanisms. The stimulators determine the "boundary condition" of the model. For example, in lysine metabolism, the level of the internal lysine metabolite is influenced by lysine transport into the cell, by the yield of the lysine biosynthetic pathway, by the rate of lysine degradation, and by the rate of lysine utilization in proteins biosynthesis. The external lysine level, on the other hand, is assumed to be determined and kept fixed by the experimenter throughout the experiment.

3 Computing Modes

Given a model stimulation q we would like to compute the set of model's modes whose stimulators states coincide with those of q. This will be the first step in using a model to infer the state of the system under a certain condition.

A *q-mode* of a model M and stimulation q is a mode m such that for each $v \in U_I$, $q(v) = m(v)$. We denote the set of q-modes by $Q_{q,M}$. A model M with acyclic graph G_M is called a *simple model*. We note that q-modes are unique and easily computable for simple models: Given a stimulation q and a topological ordering on the graph's nodes (which exists, since the graph is acyclic), we can compute the q-mode by calculating the state of each variable given its regulators' states. In summary:

Claim. Let M be a simple model where $G_M = (U, A)$. For any stimulation q, there is a unique q-mode that can be computed in time $O(|U| + |A|)$.

In practice, model graphs are not acyclic and feedback loops play a central role in system functionality. In cyclic models, a stimulation q may have no q-modes (in case no steady state is induced by the stimulation), a unique q-mode, or several q-modes. In order to compute the set of q-modes we will first transform a cyclic model into a simple one. Recall that a *feedback set* in a directed graph is a set of nodes whose removal renders the graph acyclic [6]. A feedback set of a model M is a feedback set for the graph G_M. Given a feedback set F, the *auxiliary model* M_F is obtained by changing the regulation functions of the variables in F to null. The graph G_{M_F} is updated accordingly and becomes acyclic, so M_F is simple. Given a set $F' \subseteq F$, we say that a mode m' of M_F is (M, F')-*compatible* if m' agrees with M on every $v \in F'$. In particular, a mode of M_F which is (M, F)-compatible is also a mode for M, since the steady state requirements hold for every $v \in U \setminus F$ (by definition of M_F modes) and for all $v \in F$ (due to the compatibility). Given a mode for M_F, it is easy to check if it is (M, F')-compatible by calculating f_v for each $v \in F'$. The following algorithm calculates the q-modes of M by using a feedback set F and a topological ordering of G_{M_F}:

Mode Computation Algorithm
- Generate each possible state assignment to F. For the assignment $s_F : F \to S$ do the following:
    - Generate a stimulation q' for M_F by joining q and s_F.
    - Use the topological ordering to compute a (unique) q'-mode m'.
    - If m' is (M, F)-compatible, add it to $Q_{q,M}$.

Hence, we have shown:

Proposition 1. *Given a model M, a feedback set F, and a stimulation q, the q-modes can be computed in $O(k^{|F|}(|U| + |A|))$ time.*

We note that the minimum feedback set problem is NP-hard [12], but approximation algorithms are available [15]. The complexity of our algorithm is exponential in the size of the feedback set, but this is tolerable for the current models we have analyzed. Much larger systems may require heuristics that avoid the exhaustive enumeration of feedback set states we are currently using.

4   Experimental Conditions and Their Inferred Modes

An ultimate test for a model is its ability to predict correctly the outcome of biological experiments. We formally represent the data of such experiments as *conditions*. A condition e is a triplet (e_q, e_p, e_s). e_q is a model stimulation defining the environment in which the experiment was performed. e_p is a partial assignment of states to variables in $U \setminus U_I$, and is called a *perturbation*. A perturbation defines a set of variables whose regulation was kept as a particular constant during the experiment. For example, knockout experiments fix the state of mRNAs to zero. e_s is a set of measurements of the states of some variables, and is called an *observed partial state*. We define $e_s(v) = -1$ for variables that were not measured in the experiment. Low throughput experiments (like northern blot or ELISA) typically measure one or few variables in a given condition. High throughput experiments (e.g., gene expression arrays or protein expression profiles) may measure the states of all variables of a particular type. A different type of high throughput experiments are growth sensitivity mutant arrays [4]. Each such array corresponds to many conditions, all with the same stimulation (representing the environment of the experiment), but with different perturbations (different knocked-out genes), and only a single measured variable: the growth level. We will assume that this level corresponds to the yield of the metabolic pathway under study.

Given a condition e we wish to use a model M to compare the possible modes induced by the stimulation e_q with the observed partial state. If the condition involves a perturbation, we first have to update our model accordingly. For simplicity assume this is not the case. We then apply the algorithm from the previous section and compute the set of all e_q-modes. In case more than one exists, we expect the correct one to be most similar to the observed partial state. To assess this similarity we introduce a score function that equals the sum of squared differences between the observed partial state e_s and a e_q-mode. Precisely, given a condition e and an e_q-mode s, we define the discrepancy $D(s, e)$ as $\sum_{v \in U, e_s(v) \neq -1}(s(v) - e_s(v))^2$. The mode with smallest discrepancy will be considered as our *inferred mode*. Its score is called the *model discrepancy* on condition e, i.e., $D(M, e) = \min_{s \in Q_{e_q, M}} D(s, e)$. If no e_q-mode exists, $D(M, e)$ is set to a large constant K. Note that models with loosely defined regulation functions may have a large number of modes per stimulation and consequently suffer from over-fitting of the inference.

5 Learning Regulation Functions

Given a model and experimental conditions, we wish to optimize one particular regulation function in the model and in this way derive an improved model with lower discrepancy. In this section we discuss the resulting function optimization problem, and show that this problem is NP-hard. We translate the function optimization problem to a combinatorial problem on matrices, and provide a polynomial-time greedy algorithm for it. Finally, we show that the greedy algorithm guarantees a $1/2$-approximation for a maximization variant of the function optimization problem.

We focus on one model variable v and fix the set of v's regulators $\{r_v^1, ...r_v^{d_v}\}$. Let $E = \{e^i\}$ be the set of experimental conditions. In order to simplify the presentation, we assume throughout this section that experimental conditions have empty perturbation sets. Given a function $g : S^{d_v} \to S$ we define $M(g, v)$ to be the model M with the single change that $f_v = g$. The *discrepancy score of g* is defined as $\sum_i D(M(g, v), e^i)$.

Problem 1. **The function optimization problem**. The problem is defined with respect to a model M, a set of conditions E and a variable $v \in U$. The goal is to find a regulation function $f_v = g$ with an optimal discrepancy score. In other words, we wish to compute $argmin_g \sum_i D(M(g, v), e^i)$.

In most extant gene networks models [9, 3, 16], an optimal regulation function can be easily learned given the topology of the network. This is done using the multiplicities (or probabilities) of different combinations of observed states for the regulators and regulatee. The main difficulty with our version of the learning problem is that the states of regulators are frequently not observed, and have to be inferred together with the regulation function. A naive algorithm can test all $k^{k^{d_v}}$ functions for the best discrepancy, but this strategy is impractical even for modest k and d_v ($3^{3^3} > 10^{12}$). In fact, the optimization problem is NP-hard (we omit the proof here).

Proposition 2. *The function optimization problem is NP hard.*

We shall translate the function optimization problem to a combinatorial problem on matrices and develop an approximation algorithm to solve it. First, we define an auxiliary matrix and show how to construct it. We define $Q^v_{q,M}$ as the set of model states s which satisfy for all $u \in U_I$, $s(u) = q(u)$ and agree with M on all $u \in U \setminus U_I, u \neq v$. Note that $Q^v_{q,M}$ is a superset of the set of q-modes $Q_{q,M}$ in which we relax the requirement for agreement on v. Given an instance of the learning problem, we form a matrix W^v with a column for each condition and a row for each assignment of states to v and its regulators. Let $\bar{r} = (r^1_v, \ldots r^{d_v}_v)$, $\bar{x} = (x_1, \ldots, x_{d_v})$. We define the matrix entry $w^v_{i,((x_1,\ldots,x_{d_v}),x)}$ as $\min\{D(s, e^i_s) | s \in Q^v_{e^i_q,M}, s(\bar{r}) = \bar{x}, s(v) = x\}$ or a large constant K if the minimization set is empty. In the following algorithm, we show how to compute W^v by relaxing the requirement for v compatibility in the mode computation algorithm. Later we shall show how to use W^v to compute the discrepancy score.

Matrix Construction Algorithm
- Initialize all entries in W^v to K.
- Form a feedback set F such that $v \in F$.
- For each condition i and for each assignment s_F of states of the feedback set do:
  - generate a stimulation q' for M_F by joining e^i_q and s_F.
  - use a topological ordering on G_{M_F} to compute a (unique) q'-mode m' for M_F.
  - If m' is $(M, F \setminus v)$-compatible, compute its discrepancy x.
  - Replace the entry $w^v_{i,((m'(r^1_v),\ldots,m'(r^{d_v}_v)),m'(v))}$ by x if the latter is smaller.

Lemma 1. *Given a model M, a set of conditions E and a feedback set F such that $v \in F$, the Matrix Construction Algorithm correctly computes the matrix W^v in*

$$O(k^{d_v+1}|E| + k^{|F|}(|U| + |A|)|E|).$$

Proof. Matrix entries are computed by minimization of discrepancies over all $(M, F \setminus v)$-compatible modes that have a given regulator/regulatee states. But $(M, F \setminus v)$-compatible modes are exactly the modes in $Q^v_{e^i_q,M}$ which are used in W^v's definition. Therefore, the algorithm correctly computes W^v. The algorithm spends $O(k^{d_v+1}|E|)$ (the size of W^v) time in initialization and $O(k^{|F|}(|U| + |A|)|E|)$ time to compute all mode discrepancies.

Lemma 2. *The discrepancy score of a regulation function g equals*

$$\sum_{i=1}^{|E|} \min_{\bar{x} \in S^{d_v}} w^v_{i,(\bar{x},g(\bar{x}))}.$$

By the last lemma, the scores of all possible regulation functions can be derived from the matrix W^v. To find the optimal function we first translate the problem to the following combinatorial problem:

Problem 2. **The Rows Subset Cover Problem**. We are given a non-negative integer valued $n \times m$ matrix W and a partition of the rows to disjoint subsets $B_1, \ldots, B_l$. A *row subset R* is a set of rows $b_1^R \in B_1, b_2^R \in B_2, \ldots, b_l^R \in B_l$. Our goal is to find a row subset with maximal score $c(R) = \sum_{j=1}^m \max_{i=1}^l w_{b_i^R, j}$.

In our settings, rows are pairs $(\overline{x}, x)$ and columns are conditions. The subsets B_j are the sets of rows with identical regulator states $\overline{x}$. To formulate the function optimization problem as a row subset cover problem we rewrite $w_{ij} = K - w_{ij}^v$. A selection of $b_i = (\overline{x}, x)$ corresponds to the setting of $f_v(\overline{x}) = x$.

The previous discussion implies that for constant value of d_v and k, the row subset cover problem is NP-hard. A *Greedy Row Subset Algorithm* applies naturally to this problem: We start with an arbitrary row subset S, and repeatedly substitute a row to improve the score, i.e., setting $S \leftarrow (S \setminus \{b_i^S\}) \cup \{b_i'\}$ where $b_i' \in B_i$ and the new S has improved score. The algorithm terminates in a local optimum when no single row substitution can improve the score. Since the score increases at each iteration and all scores are integers bounded by K, the greedy algorithm will terminate after $O(nmK)$ steps. For the function optimization problem, $O(|E||U|k^2)$ is an upper bound on the maximal score and hence on the number of steps. Each step costs $O(|E|k^{d_v+1})$ in order to find an improving substitution, and thus the total cost is $O(|E|^2|U|k^{d_v+3})$.

Proposition 3. *The greedy algorithm guarantees a 1/2-approximation for the Row Subset Cover Problem.*

We omit the proof here. Note that in practice, we find regulation functions by executing the matrix construction algorithm and applying the greedy algorithm to the obtained matrix. In order to take condition perturbations into account, we have to consider a slightly different model in each condition. For example, if a condition was measured in a strain knocked-out for a specific gene v, we will form a modified model with altered (constant) f_v function and compute its modes and discrepancy as described above. The other algorithms (matrix generation and row selection) remain unchanged.

6   Results

We applied the *MetaReg* modeling scheme and algorithms to study lysine biosynthesis in the yeast *S. cerevisiae*. This system was selected since a) it is a relatively simple metabolic pathway, b) its regulatory mechanisms are relatively well understood, and c) several high throughput datasets which include experimental information pertinent to lysine biosynthesis are available.

6.1   A Model for Lysine Biosynthesis

We have performed an extensive literature survey and constructed a detailed model for lysine biosynthesis and related regulatory mechanisms. Lysine, an essential amino acid, is synthesized in *S. cerevisiae* from α-ketoglutarate via homocytrate and α-aminoadipate semialdehyde (αAASA) in a linear pathway in which eight catalyzing enzymes are involved. The production of lysine is controlled by several known mechanisms:

(1) Control of enzymes transcription via the general regulatory pathway of amino acids biosynthesis. Starvation for amino acids, purines and glucose, induce the synthesis of GCN4ap[1] which is a transcriptional activator of enzymes catalyzing amino acids biosynthesis in several pathways, including lysine. GCN4ap is controlled on the translation level by the translation initiation machinery. Specifically, GCN2ap (a translation initiation factor 2α kinase) is known to mediate the de-repression of GCN4m translation in nutrient-starved cells. The activity of GCN2ap is induced by high levels of uncharged tRNA under starvation conditions [5].

(2) Transcription control of several catalyzing enzymes is regulated by αAASA. The control is mediated by the LYS14ap transcriptional activator in the presence of αAASA, an intermediate of the pathway acting as a coinducer. αAASA serves as a sensor of lysine production [13].

(3) Feedback inhibition of homocytrate synthase isoenzymes (LYS20ap and LYS21ap) by lysine. The first step of the lysine biosynthetic pathway is catalyzed by LYS20ap and LYS21ap. At high levels of lysine, LYS20ap and LYS21ap are inhibited, and thus the production of the pathway intermediates and of lysine itself is reduced [8].

(4) MKS1ap down-regulates CIT2m expression and hence cytrate-synthase production which is needed for the synthesis of α-ketoglutarate. The resulting limitation of α-ketoglutarate decreases the rate of lysine synthesis. MKS1ap is activated in nutrient-starved cells [7, 18].

   In Figure 2, we present the model graph of lysine biosynthesis as described above. The graph includes the lysine biosynthetic pathway, the catalyzing enzymes and their transcription control, and the translation initiation machinery controlling GCN4ap state. The model includes also external amino acids and ammonium (NH3). These are needed as stimulators to represent the environmental conditions enforced on the system. The transport of amino acids and ammonium into the cell is facilitated via specific permeases, and the level of internal amino acids and ammonium is determined by the extracellular metabolites and by the activity of these permeases. The state of internal lysine depends on the lysine transport into the cell and on the yield of the lysine biosynthetic pathway. Note that in order to study the model in relative isolation from other pathways and regulatory systems, we had to exclude some of the known relations (e.g., CIT2 and the Kreb cycle in α-ketoglutarate production, tRNAs in GCN2ap activation). The model graph contains several cycles that correspond to three distinct feedback cycles: general nitrogen control regulation (e.g. GCN2ap → GCN4ap → LYS1,9m → LYS1,9ap → ILys → GCN2ap), lysin negative regulation (LYS20ap/LYS21ap → IHomoCytrate → αAASA → ILys → LYS20ap/LYS21ap) and αAASA positive regulation (e.g. LYS14ap → LYS2m → LYS2ap → αAASA → LYS14ap). We used a feedback set *F* consisting of *GCN2ap* and IαAASA in all the computations reported below. The complete

---

[1] We use variable affixes to indicate types. m suffix: mRNA, ap suffix: active protein. Metabolites names are prefixed to indicate their type, I: internal, E: external.

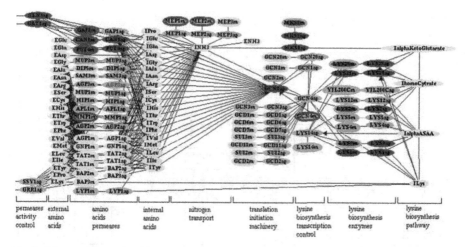

Fig. 2. The model graph of lysine biosynthesis in S. cerevisiae. Variables are represented by nodes. Arcs lead from each regulator to its regulatees. All arc directions are at any angle to the right or straight down, unless otherwise indicated. The model includes also a regulation function for each regulated variable. These functions are not shown here. Node colors indicate the mode inferred states and the observed states in condition of nitrogen depletion after 2 days. Internal node color: inferred state. Node boundaries: observed state. Red(dark): state= 2. Dark pink(grey): 1. Light pink(light grey): 0. The representation enables us to view the disagreements as color contrasts between the observed and inferred states. For example, LYS9m (bottom right) inferred state is 2 while its observed state is 1.

and annotated list of regulation functions that are part of the model, is available upon request.

We used the state space $S = \{0, 1, 2\}$. In our experiments, the definition of compatibility used for the calculation of q-modes was relaxed a bit to include also cases where $m'(v)$ and $f_v(m'(r_v^1), \ldots, m'(r_v^{d_v}))$ are both non-zeros (i.e., cases where inferred state was 1 and observation 2 or vice versa are not considered violation of compatibility). In other words, $D(i, j)$ was $(i - j)^2$ for all states $\{i, j\} \neq \{1, 2\}$, but $D(1, 2)$ and $D(2, 1)$ were set to 0. This was done to allow more flexibility in the model and to focus more on major discrepancies.

6.2 Data Preparation

We formed a heterogeneous dataset from five different high-throughput experiments: (a) 10 expression profiles in nitrogen depletion medium after 0.5h, 1h, 2h, 4h, 8h, 12h, 1d, 2d, 3d, 5d of incubation [10]. (b) 5 expression profiles in amino acid starvation after 0.5h, 1h, 2h, 4h, 6h of incubation [10]. (c) 10 microarray experiments of His and Leu starvations and various GCN4 perturbations [5]. (d) protein and mRNA profiles of wild type strain in YPD and minimal media [19]. (e) 80 Growth sensitivity phenotypes [4]. The growth phenotypes were measured for each of a collection of ten gene-deletion mutant strains in eight conditions: Lys, Trp and Thr starvation, three minimal media and two YPG conditions.

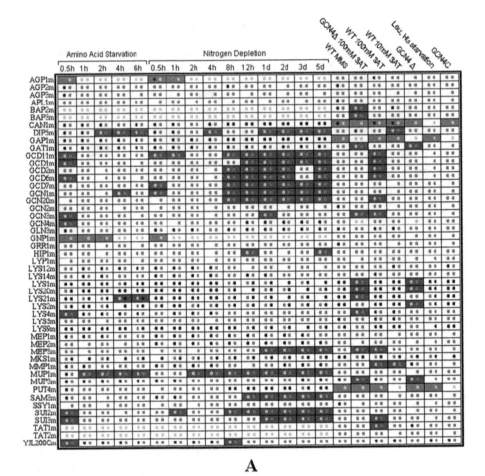

A

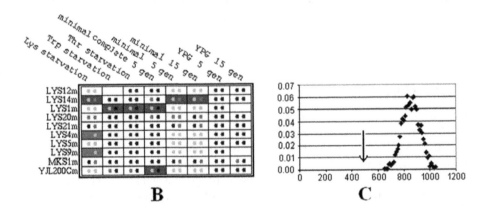

B

C

Fig. 3. Caption on page 109

To incorporate these data into our framework, we generated conditions from each of the experiments. To this end, we identified the stimulation and perturbation that match each experiment from the respective publication. We then converted the data into a set of observed states.

6.3 Model Discrepancy

For each of the high throughput conditions in (a) through (d) we computed inferred modes and compared them to the observed states. Recall that the environment defined by the condition's stimulation gives rise to a set of possible inferred modes, and we choose the inferred mode which fits the observed states best. Typically, there are only few modes per condition in the lysine model, confirming the relatively good characterization of the system by the model.

Figure 3A summarizes the comparison between inferred modes and observed states for expression conditions. Figure 3B does the same for growth sensitivity data. In general, there is good agreement between the inferred and observed states. The matrix view highlights conditions and variables in which the observations deviate from the model predictions.

Before analyzing the deviations, we verified the specificity of the total discrepancy. Since the mode computation algorithm involves selection of one mode from several possibilities in each condition, we wanted to verify that this process does not cause overfitting. To this end, we generated randomly shuffled data sets in which we swapped the states between variables of the same type. Figure 3C shows the discrepancy distribution obtained from this experiment, and supports the high specificity of the lysine model discrepancy.

We next examined the biological implication of two major deviations of the inference from the experimental data: First, the transcription of the translation initiation machinery (GCD1,2,6,7,11, GCN1,20, SUI2,3) is repressed in the later phases (8h-5d) of the nitrogen depletion experiment, and this effect is not predicted by the model. Moreover, the transcription of the ammonium permeases MEP1 and MEP2 is consistently

---

Fig. 3. Model Discrepancy *(A) Discrepancy matrix for the expression data.* Columns correspond to conditions and rows correspond to mRNA variables. Each cell contains two small squares: observed (left) and inferred (right) states of the row variable in the column condition. State colors: Cyan (light gray):0, light blue (gray):1, dark blue (black):2. The background color of the cells emphasizes critical disagreement, where the inferred state is zero and the observed state is not (green or light gray), or vice versa (red or gray). *(B) Discrepancy matrix for the phenotype data.* Each cell represents a condition, which is a combination of certain environmental nutrients and one gene deletion. Columns correspond to the nutritional environment (i.e., the medium), and rows correspond to the knocked-out variable. Each cell contain two small squares: observed (left) and inferred (right) state of the internal lysine metabolite (the ILys variable) in this condition. Colors are as in (A). *(C). Distribution of model discrepancy scores for randomly shuffled data sets.* X axis: total model discrepancy. We generated the distribution by computing model discrepancy for 50 random data sets. The discrepancy of the real data set is 494 (arrow), much lower than the minimal discrepancy measured in the shuffled sets.

activated in nitrogen depletion. To the best of our knowledge, the explanation for these observations is still unclear. However, there is some evidence for involvement of the TOR signaling pathway in the regulation of this response [2]. Second, the transcription of the lysine biosynthesis catalyzing enzymes is known to be activated by both LYS14ap and GCN4ap, but the exact combinatorial regulation function is unknown. Since they are both known to be activators, we originally modeled the regulation function of the catalyzing enzymes (LYS1,2,9,20,21) simply as the sum of LYS14ap and GCN4ap. In most catalyzing enzymes, there is a clear inference deviation in two conditions with GCN4Δ strain (Figure 3A, 3rd and 6th columns from right). In addition, the growth phenotypes of LYS14 deletion strain (Figure 3B, second row) deviate from their inferred states in all conditions with nutritional limitation of lysine. Therefore, the regulation function we originally modeled for the lysine biosynthesis catalyzing enzymes is apparently not optimal.

6.4   Learning Improved Regulation Functions

To refine our understanding of the combinatorial regulation scheme involving LYS14ap and GCN4ap we applied our learning algorithm to the regulation functions of LYS1,2,4, 5,9,12,20,21. For each one, we computed the discrepancy matrix and selected an optimal regulation function using the learning algorithm outlined in Section 5. To estimate the confidence of our learned functions we used a bootstrap procedure as follows. We generated 1000 datasets each containing a random subset of 80% of the original set of conditions. For each random dataset we recalculated the optimal regulation functions for each of the enzymes. The *confidence* of the function entry $f_v(x_1, \ldots, x_{d_v}) = y$ was defined as the fraction of times y was learned as the function value for the regulators values $x_1, \ldots, x_{d_v}$. In case of ties (several function outcomes with equal scores), we split the count among the candidate outcomes. Results are summarized in Figure 4A,B.

Based on the optimal functions, we identify two enzyme sets that share a regulatory program. The expression of genes in the first set (LYS1,9,20 and possibly LYS4 and LYS21) is dependent on the presence of both LYS14 and GCN4. Both transcription factors seems to drive the transcription of enzymes in this set linearly. The second set, including LYS5, LYS12 and YJL200C require LYS14 but not GCN4 for basal expression levels. For LYS5 it seems that GCN4 may not be a regulator at all, possibly since LYS5 is not a catalyzing enzyme in the pathway under study. We note that the combination of expression and growth phenotype information was crucial for deriving this conclusion. For example, when using expression data alone, the rows with LYS14p=0 are completely undefined.

6.5   Cross Validation

We tested the predictive quality of *MetaReg* by performing leave-one-out cross validation. For the test, we used the set of enzymes $L = \{LYS1,2,4,5,9,12,20,21m\}$ as regulatees and GCN4ap, LYS14ap, as regulators. For each variable $v \in L$ and each condition c, we optimized the regulation function of v while fixing the rest of the model and hiding the data of c. We then used the optimized model to infer the mode in condition c without using the observed value of v. Finally, we compared the inferred state of the enzyme variable to the observed one, and counted the total number of correct outcomes (or fractions of outcomes in case the inferred mode was ambiguous and several alternatives existed). Using mRNA expression data only, the accuracy derived in this procedure was 78.3% (Figure 4C).

We compared the performance of Metareg to the following alternative methods: (a) A Bayesian networks [9] with a known structure where GCN4m and LYS4m are the parents of each variable in L. We learned the local probability parameters [11] using non-informative prior. To compute the accuracy, we ran a cross validation test by learning parameters while hiding one condition at a time. The overall accuracy obtained in this procedure was 61.4%, much lower than achieved by *MetaReg*. (b) An independence model: Each regulatee in L has no regulators. We predict the probability of each regulatee outcome as the background distribution of its observations. To compute the accuracy, we ran the same procedure as in (a). The overall accuracy obtained in this

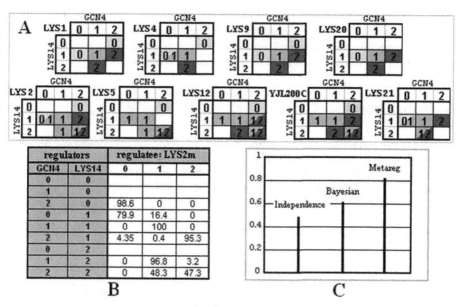

regulators		regulatee: LYS2m		
GCN4	LYS14	0	1	2
0	0			
1	0			
2	0	98.6	0	0
0	1	79.9	16.4	0
1	1	0	100	0
2	1	4.35	0.4	95.3
0	2			
1	2	0	96.8	3.2
2	2	0	48.3	47.3

B   C

Fig. 4. Learning regulation functions. (A) The optimal transcription regulation function of each lysine biosynthesis pathway enzyme as a function of the states of the regulators GCN4ap and LYS14ap. Each cell presents the state of a regulatee given the states of its regulators GCN4ap (column) and LYS14ap (row). Cell colors indicate the regulatee states. Red (dark gray): state= 2. Dark pink (gray): 1. Light pink (light gray): 0. We show only entries with over 90% confidence. For combinations of regulators states that have lower confidence or were never present in the inferred modes, we leave the corresponding entries of the optimal regulation function undefined. (B) Confidences for the LYS2 function. Rows and columns are as in (B), values are the percent of times in which the value was learned out of 1000 bootstrap experiments. (C) The accuracy of the independence, Bayesian and MetaReg methods on the lysine biosynthesis pathway. The accuracy is computed by cross validation on all expression conditions and the lysine biosynthesis pathway enzymes.

procedure was 47.5%. We conclude that the detailed modeling of interactions among proteins, metabolites and mRNAs gives an improved accuracy to our model.

7 Discussion

Models of biological regulation are becoming increasingly complex. The well established biological methodology of model development and expansion (incremental refinement) is facing major challenges with the advent of high throughput technologies and the discovery of more and more regulatory mechanisms. Computational techniques for modeling and learning biological systems are currently limited in their ability to help biologists to extend their models: De-novo reconstruction methods ignore available biological knowledge, and module-based methods do not specify concrete regulation functions. Here we aim at the construction of a computational methodology that combines well with current biological methodologies. MetaReg models can be built for almost any existing biological system, they do not assume complete knowledge of the system, and are flexible enough to integrate diverse regulatory mechanisms. Once built, the model allows easy integration of high throughput data into the analysis of the existing model. The computational tools introduced here can then be used to generate testable and easy to understand biological regulation hypotheses.

Acknowledgments

This study was supported in part by the Israel Science Foundation (Grant 309/02) and by the McDonnell Foundation. IGV was supported by a Colton fellowship. AT was supported in part by a scholarship in Complexity Science from the Yeshaia Horvitz Association.

References

1. Z. Bar-Joseph, G.K. Gerber, T.I. Lee, N.J. Rinaldi, J.Y. Yoo, F. Robert, D.B. Gordon, E. Fraenkel, T.S. Jaakkola, R.A. Young, and D.K. Gifford. Computational discovery of gene modules and regulatory networks. *Nature Biotechnology*, 21:1337–1342, 2003.
2. J. L. Crespo and M. N. Hall. Elucidating TOR signaling and rapamycin action: Lessons from *S. cerevisiae*. *Microb. Mol. Biol. Rev.*, 66:579–591, 2002.
3. P. Dhaseleer, S. Liang, and R. Somogyi. Genetic network inference: from co-expression clustering to reverse engineering. *Bioinformatics*, 16:707–726, 2000.
4. G. Giaever et al. Functional profiling of the *S. cerevisiae* genome. *Nature*, 418:387–391, 2002.
5. K. Natarajan et al. Transcriptional profiling shows that GCN4p is a master regulator of gene expression during amino acid starvation in yeast. *Mol. Cell. Biol.*, 21:4347–4368, 2001.
6. S. Even. *Graph Algorithms*. Computer Science Press, Potomac, Maryland, 1979.
7. A. Feller, F. Ramos, A. Pierard, and E. Dubois. LYS80p of *S. cerevisiae*, previously proposed as a specific repressor of LYS genes, is a pleiotropic regulatory factor identical to Mks1p. *Yeast*, 13:1337–1346, 1997.

8. A. Feller, F. Ramos, A. Pierard, and E. Dubois. In *S. cerevisiae*, feedback inhibition of homocitrate synthase isoenzymes by lysine modulates the activation of LYS gene expression by LYS14p. *Eur. J. Biochem.*, 261:163–170, 1999.

9. N. Friedman, M. Linial, I. Nachman, and D. Pe'er. Using Bayesian networks to analyze expression data. *J. Comp. Biol.*, 7:601–620, 2000.

10. A. P. Gasch et al. Genomic expression programs in the response of yeast to environmental changes. *Mol Biol Cell*, 11:4241–57, 2000.

11. D. Heckerman, D. Geiger, and D.M. Chickering. Learning baysian networks: the combination of knowledge and statistical data. Technical Report MSR-TR-94-09, Microsoft research, 1995.

12. R. M. Karp. Reducibility among combinatorial problems. In R. E. Miller and J. W. Thatcher, editors, *Complexity of Computer Computations*, pages 85–103, New York, 1972. Plenum Press.

13. F. Ramos, E. Dubois, and A. Pierard. Control of enzyme synthesis in the lysine biosynthetic pathway of *S. cerevisiae*. *Eur. J. Biochem.*, 171:171–176, 1988.

14. E. Segal, M. Shapira, A. Regev, D. Pe'er, D. Botstein, D. Koller, and N. Friedman. Module networks: identifying regulatory modules and their condition-specific regulators from gene expression data. *Nat Genet.*, 34(2):166–76, 2003.

15. P. D. Seymour. Packing directed circuits fractionally. *Combinatorica*, 15:281–288, 1995.

16. A. Tanay and R. Shamir. Computational expansion of genetic networks. *Bioinformatics*, 17:S270–S278, 2001.

17. A. Tanay, R. Sharan, M. Kupiec, and R. Shamir. Revealing modularity and organization in the yeast molecular network by integrated analysis of highly heterogeneous genome-wide data. *Proc. Natl. Acad. Soc.*, 101:2981–2986, 2004.

18. J. J. Tate, K. H. Cox, R. Rai, and T. G. Cooper. Mks1p is required for negative regulation of retrograde gene expression in *S. cerevisiae* but does not affect nitrogen catabolite repression-sensitive gene expression. *J. Biol. Chem*, 277:20477–20482, 2002.

19. M.P. Washburn. Protein pathway and complex clustering of correlated mRNA and protein expression analyses in *S. cerevisiae*. *PNAS*, 100:3107–3112, 2003.

Author Index

Lecture Notes in Bioinformatics